D1310032

DATE DUE

DEMCO, INC. 38-2931

Agricultural versus Environmental Science

A GREEN REVOLUTION

J. S. Kidd and Renee A. Kidd

CHELSEA HOUSE
PUBLISHERS
An imprint of Infobase Publishing

Agricultural versus Environmental Science: A Green Revolution

Chelsea House
An imprint of Infobase Publishing
132 West 31st Street
New York NY 10001

Library of Congress Cataloging-in-Publication Data

Kidd, J. S. (Jerry S.)
    Agricultural versus environmental science: a green revolution / J. S. Kidd and Renee A. Kidd.
        p. cm. — (Science and society)
    ISBN 0-8160-5608-0 (HC)
    1. Agriculture—Environmental aspects—Juvenile literature. 2. Environmentalism—Juvenile literature. 3. Green Revolution—Juvenile literature. I. Kidd, Renee A. II. Title. III. Science and society (Facts On File, Inc.)

    S589.75.K536 2005
    363.738'498—dc22   2004061963

Chelsea House books are available at special discounts when purchased in bulk quantities for businesses, associations, institutions, or sales promotions. Please call our Special Sales Department in New York at (212) 967-8800 or (800) 322-8755.

You can find Chelsea House on the World Wide Web at http://www.cheleahouse.com

Text design by James Scotto-Lavino
Cover design by Pehrsson Design
Illustrations by Sholto Ainslie

Printed in the United States of America

MP FOF 10 9 8 7 6 5 4 3 2 1

This book is printed on acid-free paper.

This one is for Susan and Greg

✦

Contents

Preface

The products of science and technology influence the lives of all citizens, including young adults. New means of communication and transportation, new ways of doing work and pursuing recreation, new foods and new medicines arrive almost daily. Science also engenders new ways of looking at the world and at other citizens. Likewise, science can raise concerns about moral and ethical values.

Dealing with all such changes requires some resiliency. The needed adaptations by individuals are fostered by knowledge of the inner workings of science and technology and of the researchers and engineers who do the studies and design the products. Consequently, one of the goals of the Science and Society set of books is to illuminate these subjects in a way that is both accurate and understandable.

One of the obstacles in reaching that goal is the fact that almost all the connections between citizens and scientists are impersonal. For example, the direction of study in a specialized field of science is now mainly determined by negotiations between the leaders of research projects and government officials. National elections rarely hinge on questions of science and technology. Such matters are usually relegated to secondary political status. In any case, most of the officials who are concerned with science are not elected but are appointed and are members of large government bureaucracies.

Other influences on the directions taken by science and technology come from other bureaucratic organizations, such as international political bodies, large commercial firms, academic institutions, or philanthropic foundations. However, in recent years, influence has also come from more informal voluntary groups of citizens and citizen action organizations. The scope of the set has been revised to reflect the growing importance of such channels linking citizens to the leaders in science.

The books describe some of the dramatic adventures on the part of the people who do scientific work, show some of the human side of science, and convey the idea that scientists experience the same kinds of day-to-day frustrations as does everyone.

The revisions attempt to show some of the developing trends in the impact of science on sections of the citizenry such as groupings by age or gender—or geographic location. An example is the change in the living conditions in small, rural communities that have come about as a consequence of agricultural mechanization. Finally, the books describe some of the significant strides in the actual findings of science in recent years. Some fields of science such as genetics and molecular biology have gone through a virtual revolution. These radical changes are ongoing. Likewise, the development of natural medicines was recently given social prominence by the establishment of government agencies devoted explicitly to the support of such research.

Science and Society shows the extent to which individuals can have a stake in the enterprise called science and technology—how they can cope with the societal changes entailed and how they can exert some personal influence on what is happening.

Acknowledgments

We thank the administrators, faculty, and staff of two organizations devoted to higher education—the College of Library and Information Services of the University of Maryland, College Park, and the Maryland College of Art and Design. In particular, Dean Ann Prentice and Associate Dean Diane Barlow at College Park have been extraordinarily patient and supportive.

We are also grateful for support and guidance from colleagues at the National Academy of Sciences/National Research Council in Washington, D.C. Again, special thanks to Anne Mavor and Alexandra Wigdor for their kindly dispositions and to Susan McCutchen for her high spirits.

Final thanks go to Frank K. Darmstadt, executive editor, and the rest of the staff for their dedication to the publication of this volume.

Introduction

Shades of Green, the first edition of this book, described the science supporting two concurrent but conflicting achievements; one was the development of high-yield grain, an accomplishment known as the Green Revolution; the second development was the specification of a profound environmental hazard involving the intensive use of agricultural chemicals. One result of the discord was a ban on the use of the insecticide DDT, in 1972. An accommodating response by agricultural scientists was their introduction of a program of natural pest controls called integrated pest management.

Agricultural versus Environmental Science, the revised edition, articulates how the discourse between agricultural scientists and environmental scientists has broadened. Discussions now include the issue of the mechanization of farming. A new chapter has been included as a means to provide an historical background for understanding an aspect of agricultural science and technology that appears to pit productive efficiency against the maintenance of a desirable way of life built around the family farm.

The central discourse has also broadened in other ways. After Rachel Carson opened the door to the connection between the science of ecology and environmental concerns, the connection was strengthened by the efforts of conservationists. Two new chapters are devoted to a panoramic view of the growth of ecological science and to the early linkages between ecology and agriculture. Ecological science provides a possible basis for a constructive reconciliation between environmentalists and agriculturalists.

The topic of genetic engineering has raised new issues in the central discourse. Agricultural scientists are using the techniques of genetic engineering to work toward improvements in both the quality and quantity of agricultural production. Similarly, they are seeking ways to give crops a natural immunity to pests while avoiding the use of chemical pest controls. However, some environmental scientists see risks of genetic contamination in these efforts.

Still another branch of the central discourse is the exportation of scientific agriculture to countries where there is little history of industrialization. The proponents of high-yield crops see these activities as relieving food shortages—and even famine. Some environmental scientists see a threat to the survival of family farms in these traditional cultures. The organizational structure that supports the internationalization of scientific agriculture is the topic of another new chapter.

Many specific research programs also have been undertaken since the first edition was written. Research advances in both environmental science and agricultural science are described in new chapters.

While some discord still exists between the two branches of science, substantial reconciliation has taken place over the past few years. Specifically, for example, most agricultural scientists now concede that the use of any and all means to achieve high crop yields might not be completely advantageous in the long run. New consideration is being given to ecologically appropriate practices and to preventing the depletion of our agricultural resources. A new watchword for both sides is *sustainable* agriculture.

Finally, *Agricultural versus Environmental Science* emphasizes the methods by which ordinary citizens can have a greater impact on both the private sector's and the government's strategies for the management of science. Over the past five years, events indicate that citizens' views helped change the focus of the research and development carried out by both the U.S. Department of Agriculture and by the Environmental Protection Agency. Likewise, citizen actions have influenced the direction of the activities of organizations such as the World Bank in their efforts to improve agriculture and living conditions in the developing nations of the world.

1

Confrontation

Green is the color of cultivated fields and the untouched wilderness. Both are worth protecting. In recent years, disputes have arisen about the best ways of doing so. Like opposing armies, the agriculturalists who cultivate the fields and the environmentalists who cherish the wilderness have sometimes confronted one another. Both sides see themselves as fighting under a green banner. However, the green banner of the farmer and the green banner of the environmentalist have not always been the same shade of green.

Indeed, disagreements between farmers and environmentalists have a long history. The methods that farmers use to protect crops are sometimes a source of danger to the wilderness and its wild creatures. The regulations sought by environmentalists are sometimes seen as a threat to crop productivity and to farmers' livelihood. This struggle has been affected by the many scientific advances that influence farming methods and the ways in which some people wish to protect the planet. Indeed, the scientific advances have been so profound that the process has been called a green revolution by both sides. The advances in grain productivity have been labeled a green revolution by advocates of agricultural technology. Advances in the controls over environmental hazards have been heralded as a green revolution by advocates of scientific wildlife management.

Scientists have been deeply involved on both sides of the struggle. Some of these scientists are experts in plant breeding and botany, some in preventive medicine, and others in the life patterns of birds. Indeed, dozens of scientific specialties are mobilized to advance the green revolutions.

Nonscientists are also involved. Local, state, and national officials and managers of philanthropic, academic, and commercial organizations have played leading roles. In addition, ordinary citizens of the United States and those from many countries have taken part in some version of a green revolution.

Few advances in these revolutions have been achieved in a smooth and friendly manner. The gains made by one group are often seen as losses to the other. The pursuit of legitimate goals has led to conflicts among the various parties. Scientists have played parts on both sides. In fact, some individual scientists have been aligned with different contending groups in successive time periods.

While the scientific community has been deeply involved in these controversies, the issues are also political in nature. Scientific findings on issues of public interest can be incomplete or ambiguous. The resulting disputes are difficult to resolve in ways that benefit agriculturalists, environmentalists, and the public at large.

Science and Environmental Policy

The fate of agricultural chemicals—such as DDT (dichlorodiphenyltrichloroethane)—provides a good example of the relationship between science and public policies. The 1972 banning of DDT reveals the workings of several influential institutions that use scientific findings in their planning and decision making.

On one side of the arguments about agricultural chemicals are scientists who have developed techniques for increasing agricultural productivity. The use of agricultural chemicals, such as fertilizers and insecticides, has been one of the keys to

One Effect of DDT

DDT is sprayed on crops.

DDT is ingested by insects that feed on the sprayed crops.

DDT is then ingested by birds that feed on those insects.

Birds of prey also take in the DDT because they eat the insect-eating birds.

The eggs laid by the birds of prey that have taken in DDT through their food source have thin shells and break before their young are ready to hatch.

After World War II, there was widespread use of DDT—both on farm crops and on vegetation in residential neighborhoods. In the late 1950s, science writer Rachel Carson brought to the public attention the fact that this spraying was also resulting in the death of songbirds, bringing about a "silent spring." Her book by that title, published in 1962, sparked a public battle over the use of agricultural chemicals and insecticides, leading to DDT's banning in 1972.

their success. These agricultural scientists tend to agree with predictions that expanding world populations and outdated farming practices are likely to lead to widespread famine—particularly in underdeveloped areas. Such people believe that famine and malnutrition are among the most serious problems in the world today. They view themselves as fighters in the battle against hunger.

On the other side are scientists whose efforts have been focused on matters of environmental protection. They regard themselves as fighters against health hazards such as toxic agricultural chemicals. These environmental scientists fear that cancers and other conditions can be caused by the uncontrolled use of pesticides and other chemical products. The environmentalists are deeply concerned that the misuse of agricultural chemicals will cause irreversible harm to the plant and animal populations.

Fortunately, neither side's dire predictions have been realized. Yet many problems remain. These problems are just as difficult to tackle as the scientific and social issues that had set the green revolutions in motion. Severe food shortages continue because of the increase in population, social unrest, depletion of soil fertility, and the unpredictability of the weather. Environmentalists are still distressed because there is evidence that agricultural chemicals that cause cancer and other ills can also affect animal and human reproductive processes.

Fortunately, both scientists and politicians are aware that there are acceptable solutions to these pressing problems. New technologies are becoming available. A new generation of research scientists is testing ideas that could support and improve food production without posing hazards to humans or other creatures. Of even greater importance to the protection of the environment is the recognition that all life is interdependent. This viewpoint is represented by the study of ecology.

The science of ecology describes how plants and animals—including humans—interact with one another. It defines how a community that includes both plant and animal life adapts to changing conditions. Ecology is a natural meeting ground for a diverse set of scientific specialties. Of major importance is the discipline of biology, including genetics, embryology, zoology, and botany. Chemistry also plays a major role. The study of natural chemicals—those produced by plants and animals—describes the ways in which these chemicals affect other creatures. In addition, ecology brings together the two areas of study that are central to the story—agricultural science and environmental science.

Scientists who represent these disciplines and politicians who have broader concerns hope to achieve a synthesis of the best solutions to human health and nutrition problems. They see the possibility that the two shades of green can soon be harmoniously blended.

The Making of Early Environmentalists

2

It was early morning, and the young couple lay sound asleep in their garage apartment. They were awakened by a deep rumbling sound and a throaty but muted roar. Since it was late spring, the window of their two-room flat was open, and they peered out to find the source of the noise. On the street of the posh Chicago suburb, they could see moving lights and a cloud of vapor. As the ghastly apparition approached, the sounds became louder, and they could see figures in the billowing fog.

The creatures moved closer and closer. It was soon apparent that each strangely suited and masked figure carried a long tube that extended from a hose. Each hose was attached to a noisy pump, and the end of each tube was spewing a misty gas that smelled slightly putrid.

As the procession passed in front of the garage apartment, the couple saw that the figures were directing the misty spray into the trees and shrubs that lined the street. It gradually dawned on the sleepy but agitated pair that they were seeing an attack on the local mosquito and gypsy moth population. They quickly closed their window to block off the bad smell.

On questioning their landlord the next morning, they were reassured that no harm could come from the spraying process. The insecticide being used was DDT, which had been "proven harmless" to humans. There was one small concern in the

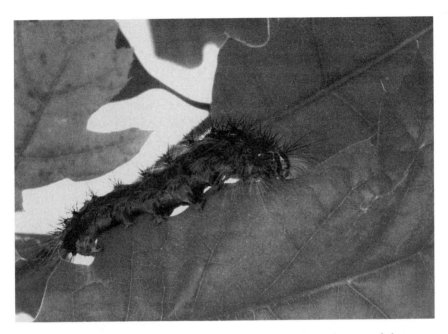

Gypsy moth grub. These caterpillars can strip hardwood trees of their leaves in a few days. After eating and going through a pupa stage, they emerge as winged moths. (Courtesy of the Agricultural Research Service, U.S. Department of Agriculture)

minds of the young couple. If the spray was harmless to humans—why were the members of the work crew wearing gas masks? No ready answer was available.

This episode took place in the late spring of 1953. It was nearly 10 years later that the hazards of DDT became a public issue. Indeed, 10 additional years passed before the U.S. federal government announced a partial ban on the use of this pesticide. The long delay was due in part to the lack of adequate scientific research. At that time, only a small amount of unbiased scientific evidence was available to the makers of public policies. Now, many more agencies are involved in ongoing studies of all aspects of environmental protection.

The Origins of the Environmental Movement

Early in the history of the United States, thoughtful citizens wished to gain both an aesthetic and a scientific understanding of the environmental conditions of their young nation. Thomas Jefferson had such goals, and might be called one of the first environmental activists in the United States. Monticello, his home near Charlottesville, Virginia, reveals his interests in the natural world and how humans can affect nature. At Monticello, he experimented with new farming techniques such as crop rotation and contour plowing. When he was president, he helped build the foundations of field biology in the United States by sponsoring the Lewis and Clark expedition of 1803 to 1806. A major goal of the expedition was to conduct an inventory of animals and plants that lived along the Missouri River and in the mountainous regions of the West.

Over the years, new voices were raised in praise of the natural environment. Writers such as Henry David Thoreau advocated the importance of protecting the natural world against the unrestricted advance of human progress. Thoreau's book *Walden*—which advanced this idea—was published in 1854.

Preservationists such as John Muir tried to convince Americans that the natural environment was a fragile asset. In the 1860s, after the Civil War, Muir began a crusade to save wilderness areas—particularly in the western United States. He believed that people should use their political power to defend the environment. At about the same time, photographers such as Timothy H. O'Sullivan and painters such as Albert Bierstadt accompanied government expeditions when they explored and mapped the West. These artists recorded the untouched beauty of the wilderness.

In contrast to such scientific and aesthetic appreciation, some people sought to exploit the opportunities provided by virgin territory. In those days, the problem of land use was fairly straightforward. Human enterprises such as farming, raising sheep or cattle, railroading, logging, mining, building factories, and establishing new cities needed land. Farmers cut down trees as if they were weeds. Aggressive developers took possession of all the land they could acquire. What they took, they changed. Lumbermen destroyed large sections of old forests by using a practice called clear cutting. In clear cutting, every tree in a designated area is cut down—whether or not it is useful for timber. The lumber companies regarded this policy as the easiest and cheapest way to move logs from the forest to the sawmill. Other developers dammed rivers and streams to make reservoirs for irrigation and later for electric power plants.

People like John Muir saw that the uncontrolled use of land was beginning to destroy beautiful wilderness areas such as the Yosemite region in California. Muir wrote the poet Ralph Waldo Emerson and other opinion leaders of his day. He urged the publishers and editors of important magazines and newspapers to mobilize public opinion toward protecting the country's natural resources. However, this was an uphill battle. Profit was usually more powerful than beauty.

In spite of many difficulties, the efforts of Muir and his followers led to some initial reforms in public policy. For

John Muir and Theodore Roosevelt. Muir's family moved from Scotland to the United States when he was 11 years old. After some years at the University of Wisconsin, he took up the life of a naturalist and spent much of his time outdoors. Roosevelt, though city bred, loved the rugged life and was a real friend to Muir. (Courtesy of the U.S. National Park Service)

Theodore Roosevelt, John Muir, and Gifford Pinchot stand with a group of forest preservationists. Together in the center, under the giant redwood tree nicknamed "Old Grizzly," are the three early environmental activists. (Courtesy of the Office of Information, U.S. Department of Agriculture)

example, the National Park Protective Act was passed by Congress in 1894. After this law was enacted, the U.S. government began to set aside public land for national parks and forests. The federal legislation was paralleled by similar laws made by the individual states.

Much of the action that followed was led by President Theodore Roosevelt. He was encouraged by Gifford Pinchot, the first American trained as a professional forester. Pinchot had studied the theory and practice of forestry in Europe, where forests were cultivated to accommodate both recreation and commercial uses. Because of this training, he was an advocate of the multiple use of land. Pinchot believed that carefully harvested forests in the United States could replenish themselves and also provide recreational resources.

During the first 50 years of the 20th century, the concern for the preservation of the natural environment continued to grow. The public expressed its interest by joining voluntary citizens' organizations such as the Sierra Club. This group was founded in 1892 by followers of John Muir and sought the preservation of wilderness areas. However, not all the voluntary citizens' groups were directed toward the same goals. Some people were interested only in recreational activities such as hunting, fishing, camping, and hiking. Others, like Thoreau, desired to contemplate the beauty of nature. A few were interested in all facets of the natural environment and were willing to transform these interests into lifelong careers.

A Complicated Life

One such person was Aldo Leopold. Over his lifetime, Leopold formed a clear philosophy about the relationship between humans and the natural world. Because of this philosophy, he is revered as a founder of applied ecology in the United States.

In 1887, Leopold was born in Burlington, Iowa, to a family of second-generation German Americans. His father was a skilled craftsperson, and his mother's father was a professional landscape architect. Burlington, on the shores of the Mississippi, was a good place for a future naturalist to spend his childhood. Game animals, birds, and fish were still abundant when Aldo was growing up. Typical of most youths of the time, he became an avid hunter of small game and an ardent fisherman. As he matured, Leopold thought of himself first as an outdoorsman and then as a naturalist.

His interest in nature prompted him to enroll in Yale University's Sheffield Scientific School in 1903. By 1906, he was ready to begin graduate studies in the recently founded Yale Forest School, endowed by the family of Gifford Pinchot. During the administration of President Theodore Roosevelt, Pinchot became the first director of the U.S. Forest Service. His family wanted to establish an American program of studies that would train professional foresters in methods unrestricted by European traditions.

When Leopold graduated with a master's degree in 1909, he immediately joined the Forest Service. He was soon made leader of a six-man expedition to map the Blue Range area of Arizona's Apache National Forest. Four members of the expedition had far more field experience than had Leopold, and at first he received little respect. However, Leopold quickly proved himself, and after only three years on the job he became the supervisor of the Carson National Forest in New Mexico.

Soon after arriving in New Mexico, he met one of the daughters of a prominent ranching family—Estella Bergere. They were married in Santa Fe, New Mexico, in October 1912. The union ultimately produced five children, three boys and two girls.

In 1913, Leopold was stricken with nephritis, a painful kidney disease. He was an invalid for 18 months and spent long hours reading and meditating on the works of prominent naturalists and nature philosophers. Co-workers believed that Leopold greatly expanded his horizons during this period of enforced idleness. After he recovered, he was intellectually and psychologically ready to accept a new challenge.

In 1915, Leopold became director of all fish and game projects in the southwestern United States. By this time, the stock of wild game and fish had been alarmingly depleted by careless or greedy people. Leopold set about restoring the game and fish populations by enforcing the local laws on hunting and fishing. He also encouraged local authorities to create reserves where wildlife could be protected and where the animals could reproduce without danger.

In that same year, the U.S. Park Service was established by the Department of the Interior. The Park Service and the Forest Service quickly became competitors for funds and public respect. The two organizations began a long debate about the proper use of federal lands. These differences included the fact that the Forest Service permitted a variety of hunting seasons on its land while the Park Service permitted no hunting of any sort. Leopold tried to bring about a compromise. He believed that the Forest Service should build game populations to provide hunters with recreational activities. While this attitude made him popular with hunters, it made him unpopular with many in the federal government and with some segments of the public.

Can the Environment Be Managed?

During the period when the United States was engaged in World War I (1917–18), Leopold served briefly as a local government official in the Southwest. He then returned to federal service. By this time, the conservation movement was gaining popularity because problems such as soil erosion and seasonal flooding were becoming more severe. Soon the Forest Service was given broader responsibility for millions of square miles of watershed land. The service sought to control the flow of streams so that rivers would be less likely to flood and wash away fertile soil. Leopold agreed with such goals.

Leopold's own plans for overall conservation were becoming more expansive and better organized. He now took into account many factors such as soil structure, steepness of terrain, and the composition of plant and animal communities within a particular territory. He saw that small adjustments in these interacting factors could produce large effects in either the retention or erosion of the soil. He recognized that effects—both positive and negative—could occur at great distances from where the initial change took place.

Leopold was also alert to the prospect that well-intentioned human actions could have negative outcomes. For example, the release of surplus horses in the western prairies had unfortunate effects when those areas became overgrazed and the top soil began to blow away. Leopold came to believe that nature acted spontaneously to repair the damage from a natural disaster such as a forest fire caused by lightning. However, human-made problems—like transplanting the horses—were longer lasting and the damage was sometimes irreversible.

While these ideas were very advanced for the time, Leopold still had some limitations. Although he provided resources to increase the population of game animals, he continued to regard hunting these wild creatures as the most important function that the natural forests could support.

Conservation or Preservation?

Leopold's superiors in the Forest Service felt that he had the ability to be a leader in the organization. By 1924, he could no longer resist the pressure to accept an administrative job in the Forest Service. He became deputy director of the Forest Products Laboratory in Madison, Wisconsin. In this position Leopold tried to improve communications between the laboratory scientists and the Forest Service people in the field. His effort had little success. He turned his energies toward traditional conservation work, such as the preservation of fish and game animals. At the same time, Leopold expanded his connections with such organizations as the Izaak Walton League, an association of people who liked to fish. He was also in touch with commercial firms that did business with hunters. These contacts resulted in feelings of mutual respect. In 1928,

Leopold left government service. Financed by the companies who equipped outdoorsmen, he become a freelance consultant on wildlife management.

This move was made possible by one of Leopold's clients, the Sporting Arms and Ammunition Manufacturer's Institute, which had promised to support his research projects for several years. The projects included studies of regional wildlife habitats and surveys of game stocks in the upper midwestern states. As he had done in the Southwest, Leopold was determined to encourage and enforce legal bag restrictions—the number of game animals that could be killed. He also hoped to persuade both public and private landowners to improve the quality of wildlife habitats on their property.

In 1931, Leopold published a handbook on the management of wildlife resources. This handbook and his many projects brought him recognition as the "father" of wildlife management in the United States.

Unfortunately, during the Great Depression of the 1930s, Leopold lost his financial support. Instead of returning to his government job, he enlarged his handbook into a college-level textbook. The text contained new, ecologically sound concepts, including the idea that good consequences can follow disastrous events such as forest fires. As an example, Leopold related that seed cones from certain pine trees develop only after being singed in a fire.

Leopold borrowed ideas from the British ecologist Charles Elton. Elton suggested that distinctive organisms can develop and thrive in each niche of a natural community. The black moth that survives in the sooty environment of certain factory towns is an example of this theory. (The moth's dark color blends in with its sooty surroundings, making it harder for predators to see and eat it.) Elton also originated the notion of the food chain—the idea that larger organisms will consume smaller, less aggressive organisms. This concept dictates the resources needed to support a given population of game animals.

The 1933 publication of Leopold's textbook generated attention from several academic institutions. Officials at the University of Wisconsin offered Leopold a professorship in their School of Agricultural Economics. State officials had become aware that much of Wisconsin's open land was in bad shape. In the northern forests private loggers had enjoyed decades of clear cutting. After the timber had been removed, the loggers abandoned their land and stopped paying taxes on it. Consequently, enormous parcels of ruined land reverted to state ownership.

State officials hoped that programs based on Leopold's ideas could restore the health of this land—quickly and relatively cheaply. Immediately after his appointment, Leopold set out to train his students to go into the field and improve the condition of the state lands.

Sand County

Leopold's teaching career proved to be a rewarding experience. By 1936, his students were beginning to fulfill the goals that had been established in 1933. However, Leopold faced new problems. The general economic depression, which had begun in 1929, had deepened in the 1930s, and the government needed to create more jobs for the unemployed. Programs such as the Works Progress Administration (WPA) and the Civilian Conservation Corp (CCC) supplied such employment opportunities.

These programs were generally successful, and many good projects resulted. However, some projects were poorly planned and violated Leopold's philosophy. For example, Leopold supported setting aside large areas of federal land to serve as game preserves or sanctuaries. The CCC, however, launched road-building projects on the same federal lands. A road into a National Forest disrupts the natural habitat and brings visitors who are often careless and destructive. The two sides, both with good intentions, were in direct conflict.

By 1935, it was increasingly evident that a systematic approach was needed to solve the problems that affected the use of public lands. Officials recognized that expanding the number of game animals while making no attempt to control this expansion was as harmful as uncontrolled hunting. Wisconsin was suffering from an oversupply of deer, and the animals were wrecking their own habitat. Soon they invaded both urban and agricultural areas and did serious damage to ornamental and crop plants. Without a sufficient food supply, many were starving to death.

At first, officials hoped that an increase in bag limits and a longer hunting season would reduce the deer population. When these ideas proved ineffective, they considered the possibility of reintroducing the deer's natural predators, such as wolves and

By working with students, Aldo Leopold was helped toward the development of a philosophy about the environment that is still influential. (Courtesy of the University Archives, University of Wisconsin at Madison)

bears. Not surprisingly, this idea met with considerable hostility. Livestock breeders saw wolves as a threat to their herds. City people declared that the idea of wolves killing deer was repulsive.

The pressing need to achieve a balance in nature caused Leopold to change his thinking. He now advocated the necessity for natural diversity in plant and animal populations rather than the arbitrary mix found in a specialized garden or park.

During World War II (1941–45), Leopold continued his teaching career. However, his workload was greatly reduced because many of his students were in the armed services or other war-related work. During this time, his family restored a vacation home on the Wisconsin River in Sand County, Wisconsin. They spent their weekends at this retreat, and Leopold was able to concentrate on his ideas of natural philosophy. He wrote a set of essays, which were published shortly after his death in April 1948. The collection of essays, published by the Oxford University Press, is entitled *A Sand County Almanac*. Leopold's book has had great cultural and political influence on the environmental movement in the United States. Many people who became environmental activists in the 1950s and 1960s were inspired by the words of Aldo Leopold.

Leopold's influence stems from his ability to represent many of the conflicting views that separated environmentalists into rival camps. While a proponent of hunting and an opponent to a sentimental approach to nature, he advocated a comprehensive view of ecology and a philosophy that valued diversity. Indeed, his plans included a place for all creatures—including the often misunderstood wild predators. His philosophy of environmental science encompassed the wide variety of attitudes that motivated people to join activist organizations.

He wrote in the *Almanac,*

We abuse land because we regard it as a commodity belonging to us. When we see land as a community to which we belong, we may begin to use it with love and respect. There is no other way for land to survive the impact of mechanized man, nor for us to reap from it the esthetic harvest it is capable, under science, of contributing to culture.

3

The Environment and Public Health

While naturalists such as John Muir and Aldo Leopold focused on problems affecting wilderness lands, other scientists were laying the foundation for the profession of public health. The connections between health and the environment have been recognized for thousands of years. In the earliest writings on health care, the Greek physician Hippocrates advised his readers to prevent illness by avoiding impure water. During Roman times, the physician Galen observed that dwelling in damp, swampy environments could cause fevers.

Through the ages, thoughtful people saw that preventing a disease was easier than curing a disease. Indeed, there have always been sets of rules or laws to help a person achieve a healthy body and mind. Many religious beliefs are indirectly tied to observations about maintaining good health. Jewish law, for example, specifically prohibits the consumption of pork. Some believe that this law was written to avoid sickness. Scientists now know that undercooked pork may contain a dangerous parasite.

A First Step

Today, cleanliness and sanitation are basic principles in all public health policies. Sanitation practices include waste water treatment, garbage collection, water purification, and street cleaning. These services are exercised on a community-wide level, while cleanliness is practiced on a personal level.

In 1864, a typhus epidemic necessitated the improvement of sanitation laws in New York City. Two years later, this emergency led to the establishment of the first municipal department of public health in the United States. The agency was called the Metropolitan Health Board of New York City.

At that time, the exact causes of disease were not well understood. However, common sense dictated that dirt and filth were associated with sickness. Everyday observations led to another important concept. Most illness was spread by contact between sick people and healthy people. Public health officials began the practice of quarantine, or isolating people with certain diseases from the rest of the population.

In 1855, Louis Pasteur, a French doctor, established the scientific basis for sanitation and quarantine policies. In that year, Pasteur demonstrated that disease could be caused by microbes. These tiny living organisms, such as bacteria, are visible only under a microscope. Microbes can easily pass from person to person through the air or by physical contact. Pasteur's discovery helped stimulate the public health movement in the United States. While it took some time for Pasteur's ideas to be understood, the dictates of both common sense and science ultimately led to the same public health practices.

In 1870, the U.S. federal government established a nationwide public health service. Although most public health concerns were local in scope, some, such as epidemics, crossed city and state borders with ease. Coordination was needed between state and local government agencies. Officials of the public health service also saw that medical research related to public health problems was too expensive to receive much support from local governments. They decided to centralize the responsibilities for a large portion of such research.

At the same time, new public health regulations required increased vigilance for maintaining purity in food and drink. For example, new laws required the inspection of food materials brought into the United States from foreign sources. In the early 1900s, the inspection of medicines was added to these regulations.

Pure Foods

As Pasteur's research revealed, foods can be contaminated with bacteria and other harmful microbes by careless handling and sloppy processing. However, food and water can also be contaminated by poisonous materials. By the 1950s, medical research revealed that certain chemicals could be major causes of cancer. The reduction or elimination of these chemicals from food and water became a matter of public health policy.

The heightened awareness of cancer-causing substances increased public concern. To calm the public's fears, Congress passed the Delaney Amendment of the Food, Drug and Cosmetic Act in 1959. This law requires that the Food and Drug Administration (FDA), now part of the U.S. Department of Health and Human Services, must prevent public exposure to any chemical that caused cancer in any animal. Even the very smallest amount of such a chemical was prohibited. Therefore, traces of insecticides on fruit would prevent the sale of the produce.

The Delaney Amendment was partly a reaction to a minor scandal. In 1957, the Department of Agriculture approved a new weed killer for use in the cranberry bogs of New England. The department recommended that the chemical be applied to the cranberry bushes *after* the fruit had been harvested. However, some growers either did not understand or ignored this instruction and sprayed the chemical on unpicked berries.

Two years later, research sponsored by the FDA revealed that this weed killer caused cancer in rats, and the agency increased its inspection of cranberries. Soon, the chemical was detected on several wholesale lots. The agency confiscated the berries and destroyed them. On November 9 of that year, the country's top health official announced the destruction of the cranberries.

The public was advised not to buy any of the fruit until all the berries on the market had been examined by the FDA. Since the Thanksgiving holiday was rapidly approaching, growers, packagers, and canners of cranberries were furious. The public was appalled. The manufacturers of the weed killer were angry. No one knew what to do. Richard M. Nixon, then vice president of the United States, attempted to suppress the panic. As reporters stood by, Nixon ate four helpings of cranberry sauce at a political dinner in Wisconsin. Nixon did not become ill, but the cranberry crisis continued for several months.

Today's average consumer knows of the dangers of ingesting cancer-causing chemicals. Since the cranberry scare, other agricultural chemicals have been found on a variety of foodstuffs. Grapes, apples, oranges, tomatoes, and wheat have all been suspect. This danger, of course, is a legitimate concern. However, there are two other important issues associated with the use of chemicals on agricultural products.

The first is the average consumer's aversion to any physical flaw on fruits and vegetables. When microbes or insects cause "bad spots" on produce, it does not sell. Consequently, farmers must prevent any sign of imperfection on their produce, but they must keep within the law by avoiding the use of questionable chemicals.

The second issue concerns the amount of agricultural chemicals that a human can safely tolerate. Scientists can detect and identify almost any chemical even in minute concentrations—as small as a few parts per billion. Researchers must now determine the level of concentration at which a chemical is truly dangerous to health.

The issue of human tolerance to various chemicals is under continuous study. However, scientists who specialize in the prevention of disease are today more concerned about contamination of foods by bacteria and other microbes than about chemical toxins. Now, most research on food purity is focused on germs rather than insecticides or other chemicals.

Insects and Disease

For several hundred years, insects have been linked to the spread of disease. In the 1500s, Mercurialis, an Italian physician, noted that flies often swarm around human excrement and then land on food that is being prepared or eaten. He speculated that something "bad" was carried by the insects to the food and then into the human body. The physician mistakenly believed that flies spread the plague known as the "black death." However, he was correct about flies being carriers of diseases of other kinds.

In 1848, Dr. Josiah Mott of Mobile, Alabama, saw a connection between yellow fever and mosquitoes. He wrote that the number of yellow fever victims seemed highest in swampy areas that were infested with swarms of mosquitoes. In 1853, an Anglo-French doctor working in the Caribbean area, Carlos Finlay, also connected mosquitoes to yellow fever. He believed that the mosquitoes carried some kind of contamination from the sick to the well. He was correct.

In the mid-1860s, existing concepts about disease transmission were forever modified. It was at that time that Louis Pasteur proved that a disease affecting silkworms was brought about by bacteria. After medical people accepted the fact that disease could be caused by microbes, they sought to discover how the microbes entered the victims' bodies.

In the 15 years after Pasteur's discoveries, the new field of bacteriology expanded rapidly. However, the mysterious transmission of some infections—especially those that involve no direct contact between victims—continued to puzzle scientists. Then in 1881, a British doctor, Patrick Manson, made a startling breakthrough while working in China. He discovered that the cause of the disabling disease elephantiasis was a tiny, microscopic worm called *Filaria*. His studies revealed that *Culex* mosquitoes carried the tiny worm in their guts. The disease was spread when a mosquito bit a victim. Manson thought that the insect introduced the *Filaria* as it sucked the human's blood. His medical colleagues scoffed at his ideas.

Patrick Manson trained in medicine in Scotland and did his research in China over a 24 year span. (Courtesy of the Photographic Archive Services, National Library of Medicine)

At that time, medical people had finally accepted the new idea that bacteria causes disease. Most, however, found it difficult to believe that a tiny worm could be a villain. Apparently, his colleagues bullied Manson to the point that he lost his temper from time to time. These emotional outbursts convinced his fellow physicians that he was mentally unstable. Nevertheless, Manson carried on with his work until the evidence proved him correct. Later, his studies revealed that the gut of the mosquito could contain other microbes in addition to the tiny worms. Manson believed that single-celled creatures called protozoa lived in the gut of the *Anopheles*

Mosquitoes have been responsible for millions of deaths from several diseases, including malaria.
(Courtesy of the Agricultural Research Service, the U.S. Department of Agriculture)

mosquito. Again, his colleagues did not believe him. However, after Manson was proven correct about the transmission of elephantiasis, he was knighted by Queen Victoria.

After Manson's success, the study of insects as disease carriers expanded rapidly. In 1889, an American, Dr. Theobold Smith, demonstrated that Texas cattle fever was caused by a one-celled protozoa that was carried by young ticks. The adult ticks carried the microbe in their reproductive glands. From there, the microbes became attached to the eggs laid by the adult female tick. As the new ticks hatched, they ingested the protozoa into their intestines. Later, when they bit a cow and sucked its blood, the young ticks passed the microbes into the cow's bloodstream.

In 1897, British scientist Ronald Ross was working in India on the widespread problem of malaria. This terrible disease infected as many as one-third of the population in many areas of that country. At first, Ross's research was regarded with the same sort of skepticism that had greeted Manson's work. However, he soon proved that the *Anopheles* mosquito was the culprit that carried malaria. Ross became a crusader for the extermination of mosquitoes in all parts of the former British Empire. Like Manson, he was knighted for his discoveries.

In the early 1900s, Walter Reed from the U.S. Army, Jesse Lazear and James Carroll from the Johns Hopkins Medical School, and Aristide Agramonte from Cuba were conducting research on the transmission of yellow fever. The men soon proved that the culprit was the *Aedes* mosquito. To accomplish this, the physicians totally isolated human volunteers from all possible sources of infection. After a period of time, they allowed each volunteer to be bitten by an *Aedes* mosquito that had previously bitten a yellow fever patient. At first, none of the volunteers became sick. The doctors had not realized that it took about two weeks for the microbes that caused yellow fever to mature in the gut of the mosquito. When they realized the problem, the doctors themselves accepted mosquito bites. In one case, the mosquito was "old." That is, it had been 12 days since it had bitten a yellow fever patient. By accident, Lazear was bitten by that particular mosquito. That mosquito bite, the only possible source of contamination, infected Lazear with yellow fever. He was the only fatality in the whole experiment.

The scientists had solved the medical mystery without ever seeing the microbe that actually caused the disease. Years later, when microscopes were much more powerful, the tiny virus was seen and identified. Such viruses are 1,000 times smaller than bacteria. They invade human cells and use the raw materials provided by the cells to make

Ronald Ross was awarded the Nobel Prize in medicine in 1902 for his studies of the transmission of malaria. (Courtesy of the Photographic Archives Service, National Library of Medicine)

American soldiers serving in Cuba were suffering from a severe outbreak of yellow fever. Together with other scientists, Walter Reed proved in 1901 that mosquitoes carried the disease. (Courtesy of the Photographic Archives Service, National Library of Medicine)

copies of themselves. Having consumed the cell's resources, they break out and seek new cells to invade.

In the 30 years between 1880 and 1910, dozens of fearful diseases—including bubonic plague (Black Death) and typhus—were proven to be carried by insects or their close relatives, such as ticks. Lice, fleas, bedbugs, and mosquitoes are all guilty of transmitting diseases that killed millions of people over the ages. Indeed, the common housefly carries a host of intestinal disorders, while the tsetse fly of Africa carries the dreaded African sleeping sickness.

It is not surprising that health care professionals despise most insects. Many believe that the total eradication of some species—like flies—would be a blessing to humankind.

Insects and Agriculture

People now know that some insects carry disease to farm animals as well as humans. However, in the business of agriculture, the problem of disease transmission is less important than the destruction of farm crops by insects. Plagues of insects and plagues of disease have been dreaded since the beginning of history. In fact, a plague of locusts is effectively dramatized in the Bible. Historians relate that North Africa has been the site of periodic insect swarms throughout the ages. Such events are not confined to the Old World. In 1848, a plague of grasshoppers almost wiped out the crops of the newly established Mormon community in Utah. As the story goes, a massive flock of seagulls swooped down on the grasshoppers and ended the threat. Gulls have been revered by the Mormons ever since.

In the 1870s, grasshopper swarms became commonplace in the midwestern farmbelt. The grasshoppers were followed in successive waves by a whole menagerie of voracious bugs and worms. After the Civil War, a vastly improved transportation system in the United States was a boon to agricultural insects. The potato beetle, native to a restricted area of Colorado, was transported back east to Illinois by unwitting travelers. In its natural habitat, the pest had its own enemies and was kept in check by limited food sources. In Illinois it found a haven in the potato fields, where it had plenty to eat and no enemies.

In addition to their migrations within the United States, insects invaded from other countries. The cotton boll weevil, for example, traveled to the United States from Central America via Mexico.

Grasshoppers consuming wheat. Grasshoppers are a variety of locusts and, while mainly a nuisance, can appear in large swarms that ruin whole areas of growing crops. (Courtesy of the Agricultural Research Service, U.S. Department of Agriculture)

By 1900, half of the 70 insects known to be harmful to agricultural crops had come from abroad. One of these new pests might arrive in an area where it could prosper and reproduce without its natural enemies. Soon large numbers of the transplanted insects would devastate the local crops. Farmers reacted with whatever weapons they could find.

The weapon of choice soon emerged. However, no one knows who invented the homemade insecticide known as Paris green. Many farmers seem to have had the same idea at the same time. For years, Paris green was a common household product. It was originally used as a pigment in paint. Because its main ingredient is arsenic, it was also used as a rat poison. Incomplete records of farm practices suggest that a mixture of Paris green and kerosene was first used against potato beetles in 1868.

By 1900, the home production of insecticides decreased because such materials were being manufactured commercially. Research on chemical insect controls was pursued in the agricultural research centers attached to colleges and universities. Soon, in many counties across the country, agricultural extension agents were talking to farmers about insect controls. The idea of chemical weapons in the war against insects had become institutionalized.

The process of organizing a war on insects was implemented by the U.S. Department of Agriculture. The Commission on Insect Control had been a part of the department since 1878. Originally, the commis-

Colorado potato beetles reduced potato fields to stubble when first released in the central and eastern parts of the United States. (Courtesy of the Office of Communication, U.S. Department of Agriculture)

sion was headed by Charles Riley. Riley had pioneered natural pest control methods by introducing ladybugs (also known as ladybird beetles or *Vedalia*) from Australia into the orange groves of California. The ladybugs attacked the little insect—called cottony scale—that was destroying the orange crop. Riley and his team of scientists saved the citrus industry in that state.

The Commission on Insect Control remained a minor part of the Department of Agriculture until 1894. In that year, Leland O. Howard became the head of the commission. Both the power and scope of the agency were

Mature lady beetles and their grubs attack aphids that infest crops such as tomato vines. (Courtesy of the Agricultural Research Service, U.S. Department of Agriculture)

greatly expanded, and it became known as the Division of Entomology. (Entomology is the study of insects.)

Howard and his colleagues believed that insects were a dire threat to the prosperity of American farmers, and they proposed an all-out war. Unfortunately, many workers in the new field of economic entomology—the study of insects that cause harm to farm crops—were not well trained in science. Some came from closely allied areas such as biology, but many had no scientific training. Later, the availability of government jobs encouraged colleges and universities to add degree programs in this new specialty.

In 1889, members of the new science of entomology founded a professional society. In addition, they took over the publication of *Insect Life,* a journal previously issued by the federal government. At about this same time, significant numbers of professional entomologists were being hired by state and local governments as well as by the federal government.

Federal Controls

Around 1900, professional entomologists instituted a national program in an attempt to control the cotton boll weevil. At first, the federal officials focused on control by natural means. They suggested that cotton should be picked as soon as it was ripe. Then the cotton plants should be plowed under or burned in the fields. In this way, the weevils would be deprived of food and unable to survive their winter hibernation. However, the farmers resisted this practice because they were accustomed to carrying out a second cotton picking in the fall. If the plants were burned or plowed under, they would miss this extra source of income.

In addition, government agents recommended that farmers plant the cotton in wider rows and keep the fields free of fallen leaves and other plant litter. These methods would remove nesting sites for the immature larvae. The agents also suggested that farmers try to rotate their crops by alternating cotton plantings with plantings of beans or corn. These alternate crops would deprive the weevils of a regular supply of cotton plants—their chosen food supply. Few farmers adopted these practices because they could not tolerate the extra work and extra expense.

Desperate, the farmers soon turned to the use of chemicals, which appeared to be the perfect solution to the problem. The farmers assumed that they could spray the fields, kill all the weevils, and then sit back until harvest time. Actually, the chemical cure did not work well. The weevils soon adapted to the chemicals and came back stronger than ever. After almost a century, cotton cultivation in the United States continues to use more chemical insecticides per acre then any other crop.

The farmers' rejection of natural methods for weevil control taught government agents an important lesson. Some farmers would not adopt radical changes in their agricultural practices unless they recognized a clear and immediate advantage. Indeed, these farmers would prefer to choose the easiest and cheapest solutions to their problems—even if these solutions were not very effective.

Since this was the case, U.S. Department of Agriculture (USDA) agents determined that the agency would supply them with the most potent chemicals that could be developed. New, powerful pest controls would allow the extension agents to enjoy the respect of the farmers and the chemical manufacturers. Thus, from the first decade of the 20th century until the 1970s, the driving force in economic entomology was to destroy all insect pests. After World War II, the most acclaimed insecticide was DDT. The use of DDT as a dusting powder during the war seemed to show that it was safe, cheap, and effective.

Specific Regulation

Although there had been few attempts to regulate pesticides until the 1970s, one effort was made as early as 1910. The Federal Insecticide Act was intended to keep fraudulent products from the public. Manufacturers were required to list the contents of the product on each label.

The U.S. government's role in the regulation process expanded somewhat in 1947. At that time, many new agricultural chemicals were being introduced. The new law, called the Federal Insecticide, Fungicide, and Rodenticide Act (FIFRA) required producers to register their pesticides with the U.S. Department of Agriculture. The main goal, again, was to protect farmers from purchasing useless products.

4

The Farmers

Archaeologists and anthropologists, who study human societies, ancient and modern, have found primitive, wild wheat growing in the area of southern Turkey and northern Iraq. They have unearthed evidence that this region, now occupied by the Kurdish people, was home to some of the earliest farmers. About 10,000 years ago, these farmers grew the first cultivated wheat.

In even earlier times, humans lived by hunting animals for meat and gathering fruit, nuts, roots, and edible seeds. The small bands of people probably had to relocate every few weeks so that they would have a constant food supply. These ancient nomads would have been overjoyed to find a field of ripe, wild wheat. Modern research has shown that such a field would supply a bountiful harvest. Indeed, in a day or two, one person could have gathered enough grain to provide food for a year. Agricultural scientists calculate that a year's supply of grain for one person would weigh more than 400 pounds (about 200 kilograms). If the group numbered several individuals, a year's supply of food would have been sizable.

Wheat, like other grains, does not spoil if kept dry. Therefore, keeping large amounts of grain would have been a reasonable idea—except for one problem. Since there were no pack animals in those days, early humans could not have carried such heavy loads from place to place. If the nomads wished to consume their abundant harvest, they had to remain in one location.

Nomadic people would have found it difficult to settle down. Most groups would have feasted for a few days, packed what they could carry, and moved on. Others, perhaps those with many old people and young children, decided not to travel for a time. The availability of food overcame their fear of staying in one place. Eventually, some of the more settled groups constructed permanent dwellings of mud brick. The remains of these dwellings have been uncovered by archaeologists.

Archaeologists and paleobotanists—scientists who study primitive plants—are greatly interested in the wheat kernels embedded in the ancient mud bricks. These kernels are different from the kernels of wild wheat found in the Kurdish territory of today. This dissimilarity suggests that prehistoric people cultivated crops of wheat rather than waiting for wild wheat to reappear each year.

Wild wheat is self-planting. Otherwise, it would not have survived without human attention. When wild wheat is ripe, the short stem that holds the kernels to the stock becomes brittle, the stem snaps, and the kernels fall to the ground and are blown about by the wind. This property—which assures self-planting—makes wild wheat difficult to harvest. The least touch causes the stem to break and the kernels to disperse. Therefore, much of the grain from wild wheat is lost to the farmer.

Paleobotanists reason that the cultivation of a higher-yield wheat crop began in an accidental manner. Sowing wheat is not a hard job; a handful of kernels is scattered on bare earth. Indeed, in some areas of the world this primitive planting method continues to be used. Scientists say that in any field of wild wheat, a few plants differ slightly from the rest. This is called natural variation. Some of the varieties have more flexible, less brittle stems and are therefore easier to harvest. Much more grain from this kind of wheat can be gathered by a farmer. Therefore, it seems logical that year after year farmers would sow more seeds from the flexible stem variety of wheat.

Gradually, the proportion of plants with flexible stems increased. In each succeeding year, the number of plants with brittle stems decreased. Finally, after many, many generations, most of the harvest was grown from the flexible stem variety. The ancient farmers had produced a truly domesticated food plant. This process of domestication was the first instance of artificial, or human-made, selective breeding.

No one knows whether the choice of wheat seed from plants with flexible stems was a lucky accident or a thoughtful decision. Did the ancient people see a connection between the parent plant and the plants that would grow from its seed? Or was it the simple reason that there were many more available kernels from that variety of wheat? Such questions can never be answered. One can only speculate about the practices of the first farmers.

More Varieties

The gradual variation in crop grains continued after the first domestic variety of wheat was developed. Two forces were indirectly responsible for the development of the additional varieties. One force was the gradual loss of soil fertility. Wild wheat draws nutrients from the soil that are automatically recycled when the plant dies in the field. With domestic wheat, a part of the nutrients is removed from the soil when the plant is harvested. Sooner or later, the cultivated soil loses fertility. As productivity declined and hunger increased, the leaders of the clan probably found it necessary to move their people to a different location. The growing conditions in the new locations were slightly different and new; better adapted varieties slowly evolved.

The second force was human fertility. The availability of a food supply and the time and energy saved by living in one place probably increased the birth rate. In addition, a dependable grain supply also helped decrease the number of infant deaths.

Consequently, a successful village would experience a modest population explosion. As the population grew, more food was needed. There might be rivalry within the group for the available foods—such as grain and game animals. Eventually, one or more of the families would decide to migrate to a new location. Again, slightly different forms of wheat would emerge.

Both less fertile soil and an increase in population caused people to leave their settlements. As farmers spread over the landscape, they encountered different climates and different types of soil. They found that some of their wheat plants would not thrive under new growing conditions. However, other wheat varieties adapted well to the new soil and climate. The most plentiful plants—from the sturdiest and best adapted wheat—provided seeds for the next year. Over many generations, these new strains of wheat would show marked differences from the original plants.

As farmers and farming methods spread into all parts of the Near East, Africa, and Europe, new and distinctive varieties of wheat continued to be established. Centuries later, there were many different varieties of wheat, each well-suited to local growing conditions. Today, such changes would be initiated by scientists in the field of agricultural technology. However, in ancient times, new varieties emerged very slowly as the farmers themselves gradually spread out and new ground was brought under cultivation.

The Spread of Agricultural Technology

Years before the Declaration of Independence in 1776, farmers in the United States organized groups to help one another. The groups sought to answer specific agricultural questions by sharing information. They discussed such issues as whether a plot of ground should be cleared for planting, partially cleared for pasture, or left alone as a source of timber and firewood.

These groups also addressed topics unrelated to farming. For example, in the early years after independence, farmers' children had few opportunities for formal education beyond the elementary school level. At that time, rural areas had little money to support high schools. Most colleges and universities were located in large cities far from farming communities. In addition, higher education was designed to train lawyers, doctors, and clergymen and provided few courses on agricultural developments. Farmers realized that their own and their children's educational needs were not being met.

In order to correct this lack of higher education, farmers sought help from the federal government. As early as 1830, state legislators proposed that President Andrew Jackson enforce the laws that required states to fund advanced education. Little action was taken because state governments had insufficient money for such purposes. There were a few exceptions, however. In 1855, the state of Michigan sponsored the creation of a college of agriculture.

In 1859, United States senator Justin Morrill of Vermont sponsored a law to give federal money to each state for agricultural education. It was vetoed by President James Buchanan. Morrill did not give up and proposed a similar bill in 1862. This law was signed by the newly elected president, Abraham Lincoln.

The previous month, the president and the Congress had addressed the concerns of farmers by the formation of the Department of Agriculture. The business of farming was gaining the recognition that it deserved.

The Morrill Act of 1862 provided small grants of money for schools of agriculture. It also provided grants of land for classroom buildings and for the cultivation of crops. Therefore, the institutions were called land grant colleges and universities. Today, in the United States and its territories, there are 72 such institutions with more than 1.5 million students.

These institutions of higher learning have gone far beyond teaching scientific methods of agriculture. They represent the world's largest and most successful collection of institutions to further higher education and scientific research.

It was not always so. At first, the concept of land grant institutions was thought to be seriously flawed. Few people were prepared to teach scientific agriculture; no textbooks were available; no programs of instruction had been devised. Worst of all, the farmers who had been enthusiastic about the concept began to have second thoughts. They reasoned that young people might learn more about agriculture by continuing to work on the family farm rather than going away to college.

To help the troubled young institutions, the U.S. Congress passed the Hatch Act of 1887. This law alotted money to the new colleges to support agricultural experiment stations. The stations maintained fields and laboratories to provide students with the means to test new farming ideas. The success of the experimental stations eventually benefited farmers by inspiring new textbooks, new curricula, and new technology such as synthetic insecticides.

However, in the beginning, the most important responsibility of those who staffed the experimental stations was to inform the working farmer of advances in agriculture. Farmers could then use or discard the ideas as they saw fit. This arrangement to exchange ideas gave the rural community leaders renewed confidence in the usefulness of land grant institutions.

Land for Sale

By the late 1800s, several large companies had purchased immense tracts of land in relatively remote areas of the United States. One British company owned a sizable tract along the Texas-Louisiana border. In 1885, the owners of that company decided to subdivide the land into medium-sized farms. They looked for a sales agent with a good reputation among American farmers and found a man named Seaman A. Knapp.

Indeed, Knapp had a fine reputation. He had begun his career in 1856 as a teacher of Latin and Greek at a boys' school in New York. In 1863, he moved to Vermont, where he was employed to reorganize and become vice president of a small woman's college. This was a demanding position, and eventually his health was affected. Three years later, he was injured in a schoolyard accident. His doctors blamed this incident on overwork and told Knapp that he must change his lifestyle. He followed their advice and moved to Iowa. Knapp began a new career managing a farm—a wedding gift from his father-in-law.

His first year as a farmer was a failure. He attempted to breed and raise sheep, but the harsh Iowa winter killed most of his flock. In 1869, Knapp went back to teaching and was appointed superintendent of the Iowa State School for the Blind. Gradually, his health returned. After six years as superintendent, he decided to return to farming. This time he chose to raise pigs instead of sheep and was much more successful.

While establishing himself as an effective farmer, Knapp retained his interest in teaching. He could not resist the desire to share agricultural information with his fellow farmers. Knapp's ambition caused him to become the part-time editor of the *Western Stock Journal*. This position led to a friendship with James Wilson, who later became secretary of agriculture under President McKinley. The work also led to a close relationship with the Wallace family of Iowa. In time, the Wallace family took over the *Western Stock Journal* and changed the name to *Wallace's Farmer*. In the 1920s, Henry A. Wallace founded Iowa's largest hybrid seed corn company. Later, Wallace served under President Franklin Delano Roosevelt as secretary of agriculture and in 1941 became vice president of the United States.

In 1879, Knapp accepted a position to teach and conduct research at the Iowa State College of Agriculture. He became president of the college in 1883. However, Knapp soon felt the need for a more adventurous way of life. He resigned from the college in 1886 when a British-owned land company offered him a position selling farm land in Louisiana. After his move, Knapp remained in contact with many farmers in Iowa. Soon he convinced some of them to follow him to Louisiana, buy a plot of land, and grow fields of rice.

When Knapp first arrived in Louisiana, many of the people living in the area did not believe the land was very fertile. Knapp set up demonstration farms to convince both old and new residents that the soil was productive. He demonstrated the most modern techniques of irrigation, planting, and harvesting. The demonstrations were impressive, and soon 25,000 new people had arrived. Local farmers began to follow Knapp's suggestions and many became quite wealthy as a result.

After this successful venture, Knapp formed his own land development company. He also founded a successful bank in Lake Charles, Louisiana. Knapp later set up a rice milling company and helped organize a group known as the Rice Association of America. In short, Knapp became as well known in the South as he had been in Iowa.

In 1898, at the age of 65, Knapp retired from his commercial activities. However, he began a new career. Knapp

This image of Seaman Knapp was struck on a medallion to honor his service to the American farmers of his day. (Courtesy of the Office of Communication, U.S. Department of Agriculture)

became an unpaid plant explorer for the U.S. Department of Agriculture. That department was now headed by his old friend from Iowa, James Wilson.

Knapp had a specific mission in mind. The rice plants that grew well in Louisiana had one unfortunate characteristic. The rice kernels often broke into little pieces when they were milled to remove the husks. Knapp went to Japan to find different varieties of rice. His adventure was a success, and he brought home 10 tons of seed. He also brought back seeds for new varieties of fruit trees, forage grass, and flowers.

In 1901, Knapp took charge of three demonstration farms that were sponsored by the U.S. Department of Agriculture. The government wanted to show southern farmers the advantages of growing more than one type of crop. At that time, a typical farm in Alabama grew only cotton. The agricultural scientists in Washington, D.C., feared that one-crop farmers were in danger.

This danger soon became clear. At the turn of the century, the cotton boll weevil invaded the South. Many farmers were ruined when this insect spoiled their entire crop.

Bankers who held mortgages on the cotton farms asked Knapp to help solve the problem. Knapp began to gather and spread information about combating the hated boll weevil. Most of his message was about adopting good, standard cultivation methods. Farmers were encouraged to practice careful seed selection, use fertilizers, and plant early so that cotton could be harvested before the weevils matured and ate the crop. Many farmers were not enthusiastic about these ideas. In order to convince skeptics about the validity of these practices, Knapp and his backers formed the first cooperative demonstration farm in 1903. On this farm, working farmers—not students or teachers—demonstrated practical methods to combat the weevil. Some farmers recognized the usefulness of the information and gradually changed their practices.

Unfortunately, the federal government could not take over and sustain the valuable program. Government officials had responded to the cotton crisis only after they realized that crop failure would ruin the economy of the entire South. Because of existing laws and limited funds, the U.S. Department of Agriculture could not continue its support after the crisis was over.

In 1904, after the peak of the boll weevil infestation, Frederick T. Gates was touring the South. He was chairman of the General Education Board,

The cotton boll weevil has probably cost U.S. farmers more crop losses than any other pest. (Courtesy of the Office of Communication, U.S. Department of Agriculture)

a part of the Rockefeller Foundation. Gates appreciated the benefits of the cooperative demonstration movement and vowed to provide Rockefeller money to support Knapp's work.

Rockefeller Foundation funds helped to expand the program beyond improving cotton farming practices. The program also reached out to young people and helped them form 4-H Clubs. Through these organizations, beginning farmers were taught modern farm practices and the importance of a mutual support system. This training often helped young men and women raise better cows and pigs than those raised by their parents.

The movement also included the means to provide extension programs for African-American farm families. Only a generation or two had passed since blacks were freed from slavery. Without the knowledge of new and better farming methods, many black farmers would continue to live in poverty. Since local governments did little to educate the farmers, the Rockefeller Foundation sponsored extension training programs for African Americans. George Washington Carver, the famous black chemist and inventor, served as an extension agent at this time and supervised others who performed this service.

Between 1905 and 1914, the program of agricultural extension education was funded by many different sources. Rockefeller money was supplemented by other private donations and eventually by the U.S. Department of Agriculture. In 1914, the Congress of the United States passed the Smith-Lever Act, which authorized direct support of extension work. The funding allowed each county to hire a cooperative extension agent who was a traveling teacher. Slowly and patiently, these agents persuaded farmers to adopt new practices and ideas about farming.

The Smith-Lever Act also financed an essential two-way communication system. The system was composed of faculty members at colleges and universities, research scientists at agricultural experiment stations, extension agents, and farm families. The agents would inform the farmers of new farming methods, technology, and ideas presented by the teachers and scientists. In turn, the farmers asked the extension agents to help with their problems. The agents sought answers from the educators and researchers and reported back with the solutions.

Knapp's ideas on extension education were the basis for a variety of domestic and international programs. These programs, which revolutionized agriculture, revealed that science could be applied to practical agricultural problems. The work of Knapp and his dedicated coworkers improved the practices of many farmers. The ideas spread throughout the country in the period leading up to World War II. Then, when great amounts of food were needed in the United States and in Allied countries, the United States was proven to be the most productive and efficient agricultural country in the world. It still is.

The Everyday Farmer

Today, most American farmers accept responsibility for the well-being of their environment. They feel a kinship with their land, their crops, and their farm animals. They also have a sense of stewardship and hope to pass on a healthier environment to future generations.

Modern farmers understand the dangers of agricultural chemicals such as pesticides. However, they know that uncontrolled pests—insects, weeds, mice, rats, and some birds—can readily reduce their harvests by half. The possibility that the product of hard work and risky investment can be ruined by vermin is intolerable to the farming community.

This devastated cornfield is the aftermath of an invasion of grasshoppers.
(Courtesy of the Office of Communication, U.S. Department of Agriculture)

In America the business of agriculture is technologically advanced. The pest control problems faced by farmers do not stem from lack of available resources. Government officials encourage farmers to test and evaluate the different methods and ideas. Each option must be judged by its costs, short-term benefits, and perhaps long-term disadvantages. Overall, farmers need to reach a balance between the defense of their crops and the defense of their environment.

5
Agricultural Mechanization

Until the beginning of the 19th century, most farmers used hand tools such as scythes or hoes for much of their fieldwork. To prepare the soil for planting, the most prosperous farmers used iron-tipped plows pulled by an ox or a team of horses. It was hard, slow work. Indeed, the pace of plowing—which was influenced by the condition of the soil—was one of the factors that limited the size of farms. You could not farm more land than you could plow in a few days.

During these early years, self-sufficiency was the aim of most farm families. They raised their own fruits and vegetables, kept chickens for eggs and Sunday dinners, pigs for pork, and a steer or two for beef. People who lived in cities— many of whom had recently moved from rural areas—often kept backyard gardens for fresh vegetables. Most city dwellers, however, needed to purchase fresh food from small stores or vendors who were supplied by local farmers. While a few farmers specialized in marketing perishable produce, most farmers planted a single commodity as a cash crop. In the southern states, cash income could come from cotton, corn, rice, or tobacco. In the northern states, corn and wheat were the main cash crops.

The Cotton Gin

In the early 1800s, the first steps toward agricultural mechanization brought many changes in farming. Oddly, the first big step occurred in the southern states where machine-based industrial production was rare. The breakthrough event was the invention of the cotton gin by Eli Whitney in 1793. Because of the vague wording of early patent regulations, Whitney had problems preventing others from stealing his ideas and building similar machines. His invention was

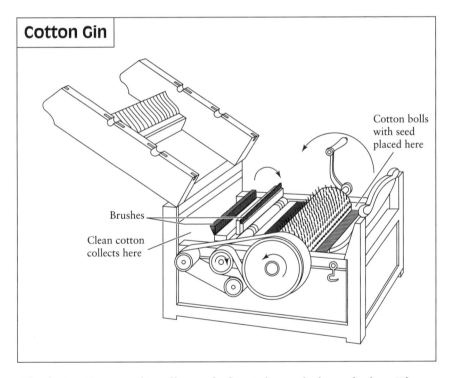

Cotton Gin

Cotton bolls with seed placed here

Brushes

Clean cotton collects here

The first cotton gins literally combed out the seeds from the lint. The clean lint was pulled through the slots between the teeth of the comb and then brushed free of the little spikes by a brush rotating in the opposite direction.

relatively simple and copies could be made by anyone who was skilled at woodworking.

To construct a cotton gin, a series of wire spikes or teeth are set in closely parallel rows on the outside of a wooden drum. The drum is mounted on an axle and rotated by a crank handle. A metal platform pierced by narrow slots is placed above the drum and positioned so that each row of teeth protrudes through one of the slots. As the drum is rotated, raw cotton is fed onto the metal platform. The protruding teeth pull wisps of raw cotton through the narrow slots. Since the seeds embedded in the cotton are too large to pass through the slots, the combed-out seeds remain on the metal platform until discarded into a tray behind the drum. The newly combed cotton is retained on the spikes until a round brush—positioned to reach the cotton-covered spikes—brushes the cotton from the spikes and into a hopper.

Prior to the advent of the cotton gin, seeds were removed by hand. It was a tiresome and boring job. Picking seeds from the long-fiber variety of cotton that grew along the Atlantic coast was difficult for the workers. However, the short-fiber cotton grown in the interior areas of the south was much, much harder to manage. The sticky seeds of this variety made it difficult to remove the seeds by hand. The cotton gin not only saved the hard work of picking out seeds but also made it profitable to grow and process the short-fiber cotton that was cultivated in the larger, interior areas of the southern states. This invention brought prosperity to a region that had suffered general poverty after the Revolutionary War.

Fortunately for the growers of the short-fiber cotton, the seeds extracted from the fibers are valuable in their own right. They have a high content of oil that can be used in cooking and in the manufacture of margarine. The remainder of the seeds, after being pressed to extract the oil, can be formed into solid cakes to make a high-protein cattle feed. Because of the cotton gin, the acreage devoted to cotton expanded rapidly, and this crop became the core element of the South's economy.

The Wheat Harvester

A few years after the invention of the cotton gin, important technical advances were being made somewhat further north. In 1843, Cyrus McCormick patented a mechanical reaper that was pulled by a single horse or mule. This machine could cut the ripe grain and save the time and manual labor required to cut the grain stalks with a scythe.

The reaper was more complicated than the cotton gin, and the design required several years to perfect. In the 1820s, McCormick's father, Robert McCormick, began work on the invention at his home in rural Virginia. The older McCormick passed his ideas on to his son around 1840. Cyrus, the oldest son, was reputed to be more of a promoter and salesman than an inventor. Nevertheless, with the help of his younger brothers, Cyrus finished the work and took out a patent. He became a door-to-door salesman in the attempt to sell the new machines near his work place. After a slow start, sales began to increase. Soon, the McCormick brothers required a steady supply of reapers—many more than they could build without additional help. Only a dedicated factory could assemble the needed number of new machines. Since the grain farms of the Midwest provided a major potential market, the McCormicks moved the center of their operations to Chicago, Illinois, and constructed their first factory there in 1848.

The early versions of the machine were built around a wooden platform with a single wheel in the rear. The front of the platform was supported by the harness attached to a horse or mule. This in-line design kept the platform from pivoting to one side. Strongly braced posts—about six feet in height—were attached near the center of the platform. A

Modern harvesters and other farm machines save both time and money.
(Courtesy of the Agricultural Research Service, U.S. Department of Agriculture)

metal axle was mounted between these posts and turned by a belt activated by the wheel as the apparatus was drawn forward by the horse or mule. Four pairs of long spokes extended from the axle and a slat-like paddle was connected across to each pair of spokes. As the axle turned, the paddles would rise to the top of their orbit and then circle down over the grain and force the stems against a cutting bar at the front edge of the wooden platform. The turning wheel not only powered the axle, but also synchronized the action so that only the proper amount of grain was cut during each rotation of the paddles. The continuing motion of the paddle after the stems were cut forced the cut stalks onto the platform so that a farm worker could rake the grain stalks off the platform to make neat rows behind the reaper. Another worker gathered the stalks into bundles—called sheaves—that were then bound with twine. The sheaves were then taken to a threshing floor where the kernels of wheat were separated from the stalks.

The growing use of machines and other forms of advanced technology greatly increased the productivity of American agriculture. Today, innovative technology has been adopted in every branch of farming. In dairy farming, for example, most cows are milked by a machine and the milk is routed directly from the cow to a refrigerated tank, with no exposure to the dusty germ-laden air.

The Tomato Harvester

An even better example of the forces that have propelled farming into the factory stage is provided by a story from the history of tomato production. The tomato plant is native to the South American country of Peru. Wild strains still grow there and produce relatively small, unattractive tomatoes. When Spanish invaders came into Central America in the mid-1500s, they found the Aztecs of Mexico cultivating a variety of domesticated tomatoes that were much more attractive than those grown in Peru. During the initial stages of the Aztec civilization, travelers or traders in the Andean regions probably had carried tomato seeds north from Peru into Mexico, where conditions supported the appearance of new varieties.

In the 16th century, Catholic missionaries accompanied Spanish soldiers into Central America. When the missionaries returned to Europe, they brought tomato seeds with them. In the mid-1600s, cultivation spread from monastery gardens in Spain and Italy into all the lands of southern Europe. Indeed, the tomato has played a vital role in the cuisine of southern Italy for many decades.

Tomato cultivation began in North America in the early 1700s with seeds brought west from Europe rather than north from Mexico. The early colonists regarded the tomato as an ornamental plant rather than a food crop. As late as 1800, Thomas Jefferson raised tomato plants as curiosities in his gardens at Monticello in Virginia. At that time, the tomato was widely regarded as a poisonous fruit. The origin of such an idea is unknown although some people may have recognized that the tomato is botanically related to the highly poisonous nightshade family. The innocent tomato perhaps suffered guilt by association.

According to legend, the tomato gained acceptance as a North American food after a dramatic demonstration in 1820. This bit of theater was performed by Colonel Robert Gibson Johnson on the steps of the county courthouse in Salem, New Jersey—a small town on the Salem River. One summer day, Johnson defied local custom by proposing to eat two fully ripe tomatoes. He had been warned by his physician not to perform this stunt, and the assembled townsfolk echoed that sentiment. Nevertheless, Johnson—who may have eaten and enjoyed the fruit when stationed in Italy—ate the tomatoes

Tomato harvester. This machine replaced the labor of dozens of migrant workers in the California tomato fields. (Courtesy of the Photographic Center, Agricultural Research Service, U.S. Department of Agriculture)

with great gusto. When he did not fall dead, everyone was convinced that the tomato was not a danger to life. Cooks gradually adopted the fruit as a wholesome food.

In spite of the wide acceptance of this story, it seems highly unlikely that such a demonstration triggered a mass movement toward tomato consumption in North America. More likely, immigrants from southern Europe introduced this food into the local cuisine, and the idea spread because of the popularity of Italian and Greek restaurants. Today, fresh tomatoes lead all produce sales in supermarkets and are the most popular backyard garden plant in the United States.

The success of the tomato results from the many ways in which the fresh or processed fruit can be eaten or drunk. Few salads are complete without some tomato wedges. Tomato juice is consumed with or without augmentation. Cooked or canned tomatoes are ingredients in hundreds of dishes. Fortunately, the tomato has high nutritional value and may include enzymes that suppress the onset of certain cancers.

The strong and growing demand for canned and processed tomatoes has encouraged expansive cultivation. For example, massive plantings are laid out in the Sacramento Valley of California, where conditions are close to ideal for tomato culture. The same area is the home of the University of California at Davis. It is not surprising that members of the School of Agriculture at UC Davis have a special interest in tomato growing. In the mid-1950s, Professor Gordie C. Hanna looked into the future and perceived that someday tomatoes would be harvested by machine. He was a botanist and plant breeder, and he set out to breed a tomato that could stand the rigor of such a harvesting device. Since tomatoes—destined to be canned or pulped for ketchup—are picked when fully ripe, Hanna needed to strengthen the physical structure of the fruit. Over several generations of various parental lines, he selected seeds from plants that produced tomatoes with higher-than-average fiber content and strong skins.

Hanna successfully bred several new varieties that were more rugged than the norm. He was just in time. In the late 1950s, Coby Lorenzen, Jr., and Steven Sluka, both on the staff at UC Davis, developed a mechanical harvester for tomatoes. This machine gathered the whole tomato plant, raked the tomato fruits from the stems, dumped the denuded vine back on the ground and scooped the fruit onto a conveyer belt where workers—standing in an attached cabin-like structure—could sort out any unripe, injured, or discolored fruits. The whole apparatus—including the cabin that enclosed the sorters—was pulled along the rows of tomato vines by a tractor.

Only a few years later, Bill Stout and S. K. Ries, two members of the research staff at the Michigan Agricultural Experiment Station, patented a similar machine. This harvester cut the stems of the tomato vines with revolving blades positioned near the ground. The vines fell upon a moving conveyer belt and were carried upward onto a second conveyer belt behind the first. This belt shook as it moved forward, and the movement separated the fruit from the vine. As in the UC Davis machine, workers standing in an attached workspace at the side of the belt could sort out any bad fruit. Finally, a third conveyer belt moved the tomatoes to a storage bin.

In both the California and Michigan cases, state officials soon made arrangements with nearby manufacturing interests, and the industrialists began to produce and market the harvesters. Adoption of the mechanical harvester by growers had some unexpected and far-reaching effects. First, the tomato harvester is unusually long. Because of the cabin-like workspace, the machine with its tractor is over 35 feet in length. This means that the ends of the tomato plots had to have wide borders so that the long machines could be turned successfully. To waste as little land as possible, farmers wanted to plant very long rows—rather than shorter rows with several machine-turning breaks. Technical advisers suggested that plantings be at least 600 feet long. This condition meant that only those farmers with very large holdings could use the machines efficiently.

A second consequence was a change in the makeup of work crews at harvest time. For the most part, men had been employed for hand harvesting. Most of these men were migrant Mexicans who came to the United States under a special program that required that they return to Mexico after the tomato harvest. The new mechanical harvesters allowed migrant men to be replaced by local women who could do the less strenuous job of sorting the fruit.

Extensive Mechanization

For thousands of years, the plow was the most important technological development in the practice of farming. This implement replaced the stick, the hoe, or other crude devices used to break up the soil. Countless early farmers must have invented many varieties of plows to help ease their labors. Roman farmers worked their fields with plows. Roman soldiers brought the technology to Britain when they conquered that land about 1,900 years ago.

In more modern times, John Deere and Leonard Andrus developed an efficient steel plow in 1838 and by mid-century, they were selling at the rate of 13,000 a year. More recently, powerful and sophisticated tractors began to replace teams of oxen, mules, or horses. Field hands have been replaced by the newer mechanical harvesters that combine reaping and threshing in a single machine. Pulverizing the soil can be done by a machine. Feed from silos is brought to farm animals by conveyer belts and similar machines. Even specialty crops such as snap beans or sweet peas can be mechanically harvested.

Progressive farmers maintain records and accounts on personal computers. They use complicated mathematical formulas to determine the size of each crop. Times of planting, cultivation, and harvesting are also aided by computer calculations. Indeed, the most advanced machinery may contain computer chips that can adjust gasoline consumption and warn of breakdowns.

As a consequence of the advances of technology, farms have become larger and larger. Such large operations must be managed as if they were factories. Ownership is so expensive that it must be shared, and the ideal form of shared ownership is the corporation. To increase their profits, large numbers of labor-saving devices have been employed by the corporations. Thousands of farm workers have lost their jobs because of these techniques. This effect is illustrated by the fact that in 1900, 40 percent of the total workforce was employed in farming. By the year 2000, that figure had dropped to 2 percent. Although the food and agriculture business is still a major source of employment, most of the workers are employed in stockyards, canning factories, business offices, and research laboratories rather than in fields and pastures. In short, farmers who work their own land are now a tiny minority of the total population.

Critics of agricultural industrialization—including many environmental activists—see the process as dehumanizing. The critics contend that the United States is moving toward monopolistic control of food production, processing, and marketing. They see the outcome of this control as strict standardization of food products. In their view, standardization is a step toward blandness and tastelessness.

These critics also believe that it is unnatural for the managers of factory farms to crowd thousands of chickens into a restricted area. They maintain that it is cruel to confine chickens to a foot of space in a massive, foul-smelling shed.

The opponents of mechanization contend that the whole complexion of rural areas has changed as the big companies buy out small farmers and put in bureaucratic management. In many cases, the local businesses that supplied the commercial needs of the smallholders have lost their customers and closed their stores. Some small towns—and their special quality of life—have been known to disappear.

Some of the critics blame the U.S. Department of Agriculture (USDA) for rural decline. It is true that for many years most agricultural research and development was sponsored by the USDA, and this work was focused on improving efficiency. That orientation often led to new versions of mechanical equipment. However, it is unfair to hold the USDA responsible for all the recent changes in rural life. In fact, this federal agency has become more sensitive to the concerns of people who are unhappy with the industrialization of farming. Lately, research subjects linked to ecology—such as soil science and entomology—have become more important. Studies are being designed to find ways to reduce the harm that agricultural innovations might inflict on the environment.

6
Corn

Eating sweet corn in July and August can be a great delight. However, buying and cleaning corn always presents certain problems. The careful shopper wants very fresh corn because sweet corn begins to lose its flavor as soon as it is picked. At least, that was the way things used to be.

Shortly after an ear of ordinary sweet corn is cut from the plant, the sugar in the kernels begins to turn into starch. The taste of corn on the cob is much less appealing when the sugar content decreases. Now, however, new varieties of sweet corn are bred to retain their sugar for a longer period. Because of these special breeding techniques, corn sent to distant markets can retain its sweet taste.

Although the problem of lost taste has been greatly reduced, the problem of removing husks and silk from an ear of corn will never go away. Everyone knows that removing the outer leaves of corn—the corn husks—and the corn silk is a chore. However, corn silk, long tubes that carry the male genetic material to the corn's egg cells, are essential to fertilize the seeds. Each corn silk fiber grows from an immature egg cell. As a silk elongates, it extends up the cob between the rows of other egg cells. Finally, each silk grows longer than the cob and emerges beyond the husks, which wrap around the ear of corn. When a male pollen grain falls onto an exposed strand of sticky corn silk, the pollen begins to grow a long, thin extension. This tiny strand penetrates the silken tube and then pushes down to the egg cell. The male genetic material is thus carried down the tube and fertilizes the corn's immature egg cell. Each egg cell then forms a kernel—a fertile seed. In short, one cannot have tasty ears of corn without the assistance of corn silk.

Wild corn is probably descended from a plant called teosinte. This early form of corn grew wild in Central America. Its ears were only about one inch (2.5 cm) long. Each of the five or six kernels on the tiny cob was covered by a tough shell that split open when the cob dried. This process allowed the seeds to scatter on the ground.

At least 10,000 years ago, the primitive wild corn that evolved from teosinte was collected by Native Americans of Central America. The wild corn also produced small ears. However, unlike the cobs of its ancient parents, each corn kernel was covered by a soft, leaflike husk. These thin husks shriveled as the ripe ears became dry and allowed the kernels to scatter and self-seed.

Around 4,000 years ago, the Native Americans began to cultivate corn plants rather than gather wild corn. At first, they had to brush each kernel out of its own husk before the corn could be cooked or ground into cornmeal.

Over time, these early farmers began to sow seeds from plants that had larger and longer cobs. Eventually, probably by accident, a plant produced one thick husk that covered the whole cob rather than a thinner husk for each separate kernel. After this change, the thick husks did not allow the kernels to disperse and self-seed. The ripe ears of corn could fall to the ground, but they would rot with the kernels still intact. Indeed, modern corn plants would not survive without human help.

Another problem affecting corn is infestation by ugly little worms called corn borers. They creep in through the top of an ear of corn, where the silk grows through the husk. These worms are being fought by modern breeding methods. Some corn varieties now repel the pests, and further selective breeding may produce corn that is free from corn borers.

Actually, sweet corn accounts for a rather small amount of the corn grown in the United States. Field corn is the major corn crop. Millions of pounds of this grain are produced each year. The size and taste of field corn is quite different from the sweet corn to which North Americans are accustomed. The ears and kernels are bigger, and the ripe kernels have much more starch.

In the United States humans consume only a small fraction of the field corn crop. Most of it is fed to cows, pigs, chickens, and other animals. However, in nonindustrialized countries most of the field corn is eaten by humans. For many families in Latin American countries, field corn is the main source of food. It is frequently ground into flour and is used to make tortillas (a flat, unleavened bread or pancake) and other dishes.

Corn is a member of the family of grasses. In most other species of grasses, the male and female organs exist together in the same flower. In corn, however, these organs are found in two different parts of the same plant. The pollen-producing male organ grows at the top of the stalk—or corn stem—and forms a set of tassels about 6 to 7 inches (15 to 17.5 cm) long. These tassels contain the pollen granules needed to fertilize the female parts of the corn plant. The female organs are found in the ears of corn that grow about halfway up the stalk.

This separation of male and female parts makes it easy for scientists and corn breeders to control the fertilization process. However, when corn is growing naturally, there is no control at all. Wind-blown pollen grains from many different plants land on the exposed corn silk and then fertilize the immature egg cells of the corn plants. Indeed, the kernels on one ear of corn can be fertilized by a dozen different pollen sources. Pure chance determines which pollen grains land on which strands of corn silk. This is known as natural crossbreeding.

Until the 1920s, farmers saved the best ears of corn from each harvest to provide seed for the next planting. However, this practice had a limited influence on the quality of the corn crop. Each seed—even those on the same cob—could have a different male parent and therefore a different set of desirable and undesirable characteristics.

The use of homegrown seed decreased after 1934 when hybrid seeds gained acceptance in the United States. Hybrid seed corn, the result of scientific research, was adopted because it increased the yield of a corn crop by 10 to 30 percent.

The First Green Revolution Begins

The agricultural green revolution began in 1906 at the Station for Experimental Evolution in Cold Spring Harbor, New York. The agent of the revolution was a young biologist named George Shull. Shull was fascinated by the accounts of Hugo de Vries, a Dutch scientist. A few years earlier, de Vries had rediscovered the research done by the Austrian monk Gregor Mendel. Mendel's work, published in 1866, concerned the control of plant heredity—especially that of pea plants. Shull resolved to expand and refine Mendel's theory of heredity by studying a variety of plants. As an assistant researcher, he studied the evening primrose and later as a project director, he focused his research on corn.

Shull sought to establish two purebred lines of corn plants—line A and line B—as Mendel had done with his peas. To accomplish this, all the kernels on each ear of corn must be fertilized by pollen from the same plant. In corn breeding this process is called selfing. To accomplish selfing, a newly formed ear of corn is covered by a paper bag, which is then sealed with a rubber band. When the plant matures, a tassel that holds the pollen grains is cut from the top of the same stalk. The bag is removed from the ear, and the tassel is shaken just above the corn silk. After the pollen falls onto the silk, the bag is replaced so that no "foreign" pollen can touch the corn silk.

Shull repeated this procedure for eight generations of corn plants. This number of generations was required to make sure that each line, A and B, became pure and had no genetic material from other corn plants. Then he crossed the two purebred lines and produced hybrid AB.

The results were strange indeed. As the selfing progressed, each generation of inbred stock looked worse than the previous generation. After eight generations, a purebred plant was short and wizened, the ears were small, and the rows of kernels were incomplete. Indeed, the adult corn plants looked very unhealthy. A farmer would not have been proud to grow a crop of purebred corn.

George Shull, an unheralded botanist, triggered a major revolution in the cultivation of corn. (Courtesy of the Carnegie Institution Archives)

In spite of their poor appearance, Shull crossed the two sickly, purebred lines (lines A and B). To do this, the mature pollen on the tassels of one line of plants was introduced onto the silk from the other line of plants. This cross-pollination achieved an astounding result. Instead of another generation of poor specimens, the crossbred corn plants grew up strong and tall. The ears were long and fat, and sometimes two or even three ears grew from the same stalk. Shull must have been surprised by the results. He had, by accident, produced the first super high-yielding corn. No one understands why the interbreeding of two weak, purebred lines produces a robust and vigorous offspring—but it does.

Across the Long Island Sound, at the Connecticut Agricultural Experimental Station, Shull's results were being confirmed by the research of Edward M. East and his students. Although their studies were on the same topic, the two men had different purposes in mind. East worked to improve the yield of corn harvests. George Shull, on the other hand, was committed to pure research. He was uninterested in the commercial applications of his discovery. As soon as Shull finished his studies of corn, he returned to his research on the primrose.

In 1915, Shull left Cold Spring Harbor and joined the faculty of Princeton University. Over the years he became a well-loved professor. His appearance reminded students of Santa Claus, and he fit the image of a jolly elf. Shull taught at Princeton until he retired in 1942.

The corn plant on the left is one that grew from a seed that had been pollinated in the natural way. The stunted corn plant on the right is the outcome of several generations of inbreeding called selfing. (Courtesy of the Carnegie Institution Archives)

Edward East and his colleagues at the Connecticut Agricultural Experimental Station continued the research on breeding corn. They hoped to make Shull's discovery of hybrid seed corn into a profitable project for both plant breeders and farmers. East encouraged one of his advanced students, Donald Jones, to focus on reducing the high cost of hybrid seeds.

The greatest expense in producing hybrid seeds was in the actual cultivation of purebred, self-pollinated lines of corn plants. Even before the lengthy process of inbreeding could begin, a breeder had the long-term investment of buying land and preparing the fields. Then came the expense of tending each individual plant. Using the most efficient growing techniques, it took at least four years to establish a purebred line. The stunted, purebred plants that resulted from the lengthy and expensive breeding technique had small ears and few kernels per ear. Consequently, few seeds were produced by each self-fertilized plant.

In addition to these problems, hybrid superiority was found to decline after the first generation of plants. This discovery meant that farmers could not cultivate a few bags of expensive hybrid seed and then expect that crop to produce hybrid seed for the next year. To enjoy the advantages of Shull's discovery, farmers had to buy, plant, and cultivate new and costly hybrid seed every year.

Jones thought he could reduce the price of hybrid seed. He started by developing four distinct lines of purebred stock rather than two. Using the same selfing method as the earlier breeders, he grew eight generations of self-pollinated corn plants—lines A, B, C, and D. Then, he crossbred line A with line B (hybrid AB) and line C with line D (hybrid CD). The first generation of plants from each cross was robust and productive.

Jones then crossed hybrid lines AB and CD to produce an equally hardy plant, the double hybrid ABCD. After the first generation, these plants, too, lost their excellent hybrid qualities. However, at the cost of only one additional growing season—to cross hybrid lines AB and CD—Jones achieved a tremendous breakthrough. Since the ears of both AB and CD were larger and contained more kernels, the last cross—producing ABCD—yielded many times as many hybrid seeds. Although farmers still needed to purchase new seeds for each crop, the large number of available seeds greatly reduced the selling price. Jones's discovery, publicized in 1919, was called the double-cross process.

At first, agricultural scientists were very skeptical of Jones's research. They did not understand hybrid vigor and had a hard time grasping the double-cross method. However, some seed merchants saw that hybrid vigor was genuine even if they could not explain it. The seed companies sent their sales agents to educate ordinary farmers about the benefits of growing hybrid corn. After the market was established, the seed producers began to mass-produce hybrid seed and to open distribution centers in farm communities. Soon farmers were cultivating thousands of acres of high-yielding corn. Farmers, food processors, and the public eventually benefited from the new seed. However, it took 20 years to achieve the benefits.

Agents of Change

In 1915, a large seed-corn grower took the first critical step toward marketing hybrid seed. Eugene Funk, president of Funk Farms in Bloomington, Illinois, launched a major program to improve the productivity of field corn. The company planned an extensive program to select and control crossbreeding. Earlier, they had conducted a few small tests on the methods developed by Shull and East. The tests were disappointing, and the Funks were not enthusiastic about the method.

To begin his new project, Eugene Funk hired James Holbert, a recent graduate of Purdue University. Holbert's research and his discussions with breeding experts had convinced him that hybrid corn was truly superior to any other seed. Now the young man needed to convince Eugene Funk that hybrid corn was the crop of the future. After many long discussions, Holbert's arguments finally persuaded Funk to support additional tests. Even though the experimental hybrids looked promising, many aspects of the program still needed to be worked out.

Funk's acceptance of the breeding technique was only the first step. Producing truly superior seed in quantity required costly research and development. In addition, there was no certainty that there would be a market for expensive hybrid seed corn. Eugene Funk took a risk by accepting Holbert's proposal.

One important factor helped Funk decide to accept the venture. He had been concerned that traditional corn crops were vulnerable to epidemics of corn diseases. Holbert argued that the hardier hybrid plants would be better able to ward off any such outbreaks. This argument persuaded Funk to become an ardent backer of the research.

Holbert started his project by selecting hundreds of good specimens from which to develop his purebred lines. He began cultivating the plants in 1916. In 1918, in the third year of inbreeding, a furious storm hit the test fields. All but one group of plants were ruined. It looked like a disaster.

Holbert saved the seeds from the one surviving group and continued inbreeding these plants in 1919. That year, there was no rain at all, and many plants died. It appeared to be another catastrophe. In the end, however, Holbert's efforts worked out well. The stresses of violent rain and drought had eliminated all but the strongest stock. This stock became a hardy, purebred strain that would be the parent of dozens of successful hybrid products.

The weather conditions in 1918 and 1919 had forced Holbert to delay the crossing of his purebred lines. By the time he was ready to proceed, the details of Donald Jones's new double-cross technique were known to Holbert. Indeed, his bad luck had changed to good luck.

In the years after 1920, other seed companies in the Midwest committed themselves to the new breeding methods and began to experiment with hybrid corn. However, many county extension agents and scientists associated with universities and experimental stations shared a negative attitude toward the new seeds. These people believed in conventional breeding methods. They wanted to increase crop production by improving old farming techniques. Their most influential clients—farmers who were already doing well—agreed with them. On the other hand, seed companies were working with farmers who were still struggling for success. These people were willing to attempt a new approach.

The willingness to try new strategies was also characteristic of some of the younger leaders in the business of agriculture. One such person was Henry A. Wallace. Wallace began growing prize-winning corn when he was still in high school. He soon learned that seeds from his outstanding ears of corn rarely produced high-yielding crops the following year. He was alert to the possibility that new approaches might be required to achieve significant gains in productivity. Just before he graduated from college in 1910, he read George Shull's report on purebred crosses. The next year, Wallace began his own modest corn-breeding experiments on the family farm. His interest in hybrid corn was further increased when he met Donald Jones in 1920. At that time, Jones was on an extensive tour of the Midwest, and the two men discussed the new double-cross technique.

In 1920, people at the U.S. Department of Agriculture gave Wallace some corn seeds from China. The Chinese seeds, a red kernel variety that had been inbred for two years, were crossed with one of Jones's inbred lines. The result was an outstanding hybrid. Wallace called it Copper Cross.

In that same year, Wallace's father, Henry C. Wallace, had just begun his appointment as secretary of agriculture under President Warren G. Harding. The younger man encouraged his father to persuade the U.S. government to sponsor the development of hybrid corn. The involvement of the secretary and other Washington bureaucrats in this matter disturbed workers at the experimental farms run by the Department of Agriculture. The government promotion of hybrid corn met with resistance. Most university research scientists and extension agents were convinced that the best possible seed came from champion ears of corn. They knew that each autumn the best corn was selected at county fairs and harvest festivals throughout the United States. The winning ears of corn provided seed for the next spring planting. The university people and extension agents questioned the wisdom of planting seeds from sickly, inbred plants when seed from prize-winning corn was available.

Wallace, too, was cautious about hybrid seed corn. This care was apparent in the college textbook that he wrote in 1923. He said, "What the practical outcome will be . . . no one can say. The method looks promising but there are some drawbacks." However, in this same book, he is far less cautious about another new product. He firmly stated that alcohol would someday be used for automobile fuel. He was correct, of course. Alcohol mixed with gasoline—gasohol—is used in large quantities in countries such as Brazil.

Eventually, Wallace became more confident about the future of hybrid corn. By 1925, each issue of *Wallace's Farmer* contained some statement about the benefits of the breeding technique.

Henry A. Wallace, pioneering corn breeder, was also a successful businessman and politician. (Courtesy of the U.S. National Archives Research Service)

Wallace started his own hybrid seed corn company in 1926. He went on to become secretary of agriculture under Franklin Delano Roosevelt in 1933 and became Roosevelt's vice president in 1941. In that year, the United States entered World War II as an ally of Great Britain and the Soviet Union. Fortunately, the United States had already increased its food production. Wallace deserves some credit for the expanded harvests. By the outbreak of World War II, more than 90 percent of the farms in the Midwest were planting hybrid seed. The large increase in food production needed to support the Allies was already under way. Indeed, Wallace's decision to mix science, government, and commerce had been beneficial to all.

Science and Commerce

Scientific research needs financial backing. Sometimes money used for scientific research generates a profit. Often, it does not. Even a successful product development program does not guarantee a profit.

Money for research can come from a variety of sources such as governments and private industry. The research done by George Shull was supported by money from a philanthropic foundation directed by Andrew Carnegie, a wealthy owner of steel mills and other businesses. Shull's research was motivated by the desire to reach a better basic understanding of the heredity of green plants. He had no thought that his findings would have a practical application—much less, initiate a revolution in the production of corn. In this case the practical outcome was a by-product of the search for new knowledge.

In contrast, the research of Edward East was sponsored by state and federal governments. These studies were expected to have practical applications, and they did.

The work of Shull and East exemplifies that the sponsorship and goals of research varies from case to case. Usually, basic research is conducted in universities, while practical research is carried out in the laboratories run by the government or by private industrial firms.

Agricultural research is not confined by such rules. It is conducted by many kinds of organizations: commercial, governmental, and educational. Also, the funds that support the research can come from several different sources—including philanthropic foundations. Regardless of the source, a practical outcome is usually expected from agricultural research. Any revenue from the outcome is not expected to benefit the researcher or the organization that conducted the research. However, the research must be justified by producing a clear, economic advantage. Farmers, food processors, or consumers are expected to benefit economically from the results of the research. Otherwise, it will not be supported for very long.

Shull himself did not benefit financially from his research. He was not interested in profit. However, the seed merchants were interested in financial gain. The courage and business

Hybrid corn growing in Colorado just east of the Rocky Mountains. Note that the corn stalks are not in rows but planted close together. In combination with the use of hybrid seed, this practice generates very high yields. (Courtesy of the Agricultural Research Service, U.S. Department of Agriculture)

sense of these people allowed them to invest great amounts of money to develop the market for the new, more expensive seed corn. The Funk organization and other seed merchants took many financial risks to demonstrate the advantages of their product. However, Shull's work was the first step toward profiting from the science of genetics. The research and development of hybrid corn demonstrates that money and science can support each other.

7
Wheat

When George Shull crossed two different purebred lines of corn, he produced a plant with hybrid vigor. While practical plant and animal breeding had been done for centuries, this was the first time that the pure science of genetics led directly to a practical outcome. The outcome was very impressive. Between 1920 and 1940, corn production doubled in the United States.

The development of hybrid corn was also a major success for applied agricultural science. Agronomists such as East worked cooperatively with the botanists and geneticists. Final success resulted from a collaboration between those practical scientists who advanced the agricultural technology of growing corn, technicians who applied these advances, and the business people who promoted the product to farmers.

Such success attracts attention. Scientists and others began to consider utilizing such collaboration in developing and marketing new selective breeding techniques for wheat and other grains. Many important people believed that extending selective breeding methods might be the key to fighting world hunger. Putting such an idea into action, however, was not easy.

Scientists knew that the techniques used in corn breeding could not be transferred easily to the selective breeding of other grains. Deliberate and controlled crossbreeding—vital to the hybrid process—is relatively simple in corn. The male and

The tiny stamens that produce pollen stand out from the base of each wheat flower. (Courtesy of the Office of Communication, U.S. Department of Agriculture)

female reproductive organs are separated on the plant and can be totally isolated by bagging the female parts. With wheat and other cereal grains, the male and female organs are together in each small flower. Wheat is a naturally self-fertilizing plant. Great amounts of labor are needed to crossbreed wheat.

A typical wheat plant produces 50 to 60 miniature flowers at the top of the stalk. To prevent natural self-fertilization, the male organ—stamen—of each flower must be removed. The tiny stamens are snipped out and discarded before they mature and begin to produce pollen. Then all the flowers—which contain the female organs—are encased in one small paper bag. This bagging technique, also used in corn breeding, prevents accidental fertilization from wind-blown pollen. Finally, pollen from a designated donor plant is collected. The protective bag is removed and the pollen is brushed on each flower. The wheat seeds that result from this crossing will contain a genetic contribution from each parent and will produce hybrid wheat. Unlike hybrid corn, the resulting wheat seed will breed true for many generations. Therefore, farmers can use wheat seeds from their harvest to plant the next year's crop.

Food Shortages

The United States and other technologically advanced countries have long produced surplus crops. However, many parts of the world are not so fortunate. Bangladesh, Afghanistan, Sudan, Ethiopia, and other countries have experienced serious food shortages in recent times. Natural disasters such as floods and human-made disasters such as civil war and rapid population growth have contributed to the problem. In addition, the farms in these countries often produce poor yields. Just after World War II, the average wheat crop in a nonindustrialized country yielded only about 1,000 pounds per acre—or just over one metric ton per hectare. At the same time in the United States, the average yield per acre was over 2,000 pounds, or more than 2 metric tons per hectare.

Philanthropists and scientists from the United States knew that they could not solve the problems of natural disasters, social unrest, or rural poverty. However, they could attack and possibly solve the problem of poor yields. Experience with selective breeding methods in the United States had shown that the cultivation of hybrid corn could double the yield of fertile fields. Some scientists and opinion leaders concluded that developing nations would greatly increase their food production by using the new breeding techniques with wheat.

The devastation of World War II created—or increased—harsh living conditions for people in many countries. In 1946, after the war had ended, the U.S. government undertook programs such as the Marshall Plan to help repair the war-torn, industrialized countries of Western Europe. Other government programs were begun in Japan and South Korea.

Countries that sustained little damage from active combat were given much less attention by the U.S. government. Nevertheless, administrators of philanthropic organizations such as the Rockefeller Foundation realized that some areas around the world had special difficulties. They saw local food shortages as serious, but solvable, problems.

Earlier, agricultural programs funded by the Rockefeller Foundation had been highly successful in the southern United States. These programs involved carefully focused research and a well-organized plan to educate farmers. Foundation executives believed that modern farming methods could be taught to large groups of farmers by the use of concrete examples. Instructors, employing new techniques to cultivate local fields, demonstrated how these methods increased productivity and decreased food shortages. Such demonstrations were followed with one-to-one guidance by a trained teacher. The directors hoped that by using these successful methods all farmers would adopt the new farming techniques.

Raymond Fosdick, president of the Rockefeller Foundation in the 1930s and 1940s, decided to launch a war on hunger while World War II was still under way. In the early 1940s, Fosdick consulted Henry A. Wallace, the newly elected vice president of the United States. Wallace had just returned from a goodwill tour of Mexico and was able to give an account of Mexico's farm economy. The vice president reported that the productivity of Mexican farms was poor and that more food was badly needed. Wallace believed that the use of modern technology would help improve the situation.

Fosdick decided to act on Wallace's observations. The Rockefeller Foundation had a history of successful public health projects both in the southern United States and in Latin America. Consequently, Rockefeller Foundation workers had established good relationships with many government officials in that part of the world.

In spite of these positive factors, Fosdick moved with caution. He created a survey team to study agricultural problems in Mexico. The team was led by Paul Mangelsdorf, an expert on corn breeding from Harvard University. The team also included E. C. Stakman from the University of Minnesota and Richard Bradford from Cornell University. After a year of study, the survey team endorsed Wallace's suggestions. Negotiations between the Rockefeller Foundation and the Mexican government began in 1942.

Meanwhile, the foundation

George Harrar was the scientist/ administrator who supervised the research on high-yield wheat and rice. (Courtesy of the Rockefeller Archive Center)

hired George Harrar from Washington State University to direct the overall program of applied science. Both Fosdick and Wallace recognized that there would be special difficulties in communicating with farmers from nonindustrialized countries such as Mexico. They understood the necessity of adapting the programs used to educate U.S. farmers about the advantages of selective breeding. Harrar took these problems into account as he planned a program. His projects included the education and training of local agricultural technicians. In addition to improving wheat production, Harrar's work included research on corn, beans, potatoes, and other crops, as well as the study of many aspects of Mexican agriculture. The project, located at a small agricultural college just outside Mexico City, started with the collection and cultivation of native wheat seed and comparing the yields of the different local varieties. The more sophisticated crossbreeding project started in 1944 when Norman Borlaug arrived in Mexico.

The Hunger Fighters Mobilize

Norman Borlaug was born in 1914 on a farm in northern Iowa. The strongest influence in his young life was his paternal grandfather. The older man kept young Norman by his side as he did his chores. On Saturdays he took the boy fishing and into the little town of Saude, Iowa, to buy supplies. During their days together, the grandfather taught the boy many lessons about life.

The older man counseled Norman to rely on his own decisions. In order to reach a decision, the boy was advised to gather all available information on the subject. Then he was to sift through and organize the information, apply his common sense, and make up his own mind. After reaching a careful conclusion, Norman was cautioned to stick to his decision.

Norman was also taught the importance of education. The older man saw education as the source of logical thought and the means to economic security. The boy was told that people with a good education could get and keep well-paying jobs—even in bad times.

Although the Borlaug family never had much money, they always had a pleasant home and plenty of food. In order to earn a little extra money, Norman worked at various low-paying jobs during his summer vacations. While in school, he enjoyed participating in sports and joined the wrestling team. One of his high school coaches had been on the U.S. Olympic wrestling team and encouraged Norman to learn the sport.

After he graduated from high school in 1932, Norman followed his grandfather's advice to further his education. Because of the Great Depression, finances were very tight and Norman's college choices were limited. He was accepted at the nearby Iowa State Teachers' College.

However, another opportunity arose just a week before the fall term was to begin. One of the young men from Borlaug's neighborhood was a first-string football player at the University of Minnesota. The coach at Minnesota had asked him to scout for new football players in Iowa. The neighbor recruited Borlaug and a teammate from the wrestling squad. The football player drove the new recruits to the Minnesota campus. The younger men were offered a room to share and part-time jobs to pay for their meals.

There was just one problem. Since both were from out of state, they had to take entrance examinations. Borlaug passed only part of the exam. He was required to take remedial courses during the fall term. After some difficulties with his faculty adviser, he was finally permitted to enroll in the College of Agriculture. In the spring of 1934, Borlaug began his degree program, competed in college wrestling, and worked for his meals.

One event during Borlaug's wrestling training influenced the course of his future career. Borlaug was too heavy for his normal weight class and was required to follow a rigid diet for a week. At the end of the week Borlaug was so short-tempered that he assaulted one of his own teammates. He had lost control. Later, he realized that extreme hunger can cause people to become violent. Borlaug saw this problem as an important reason to prevent famine.

Borlaug majored in forestry after his admittance to the degree program at the University of Minnesota. In the summer of 1936, he worked with the Northeastern Forest Service at Hopkins Forestry Station near Williamstown, Massachusetts. This was Borlaug's first experience with research. He supervised the planting of new trees and determined their suitability for refor-estation projects.

The next summer Borlaug was posted by the National Forest Service to operate a lookout station at Cold Mountain, Idaho. The station was 45 miles (72 km) from the nearest road, and Borlaug spent the summer in solitude. While at this remote site, he developed a lasting love for the wilderness. When the sum-mer was over, Borlaug was offered a regular, full-time job with the Idaho National Forest Authority. The job was to begin as soon as he finished his last term at the university.

Borlaug had courted Margaret Gibson since his arrival at Minnesota. Because of the job offer, he and Margaret decided to get married as soon as possible. The ceremony was held less than a week after he proposed. The young couple thought their future was determined. They were wrong.

The body louse, here magnified many times, is a notorious disease carrier. (Courtesy of the Office of Communication, U.S. Department of Agriculture)

A few weeks later, while Borlaug was working in the forestry laboratory, Professor E. C. Stakman visited the facility. Stakman was famous for his study of plant disease. He quizzed Borlaug for a few minutes and then went on his way. Borlaug's curiosity was aroused. A few days later, the young man went to hear Stakman give a special lecture on plant disease caused by fungus. The older man showed how the prevention of plant disease could reduce the likelihood of famine. Borlaug was inspired when Stakman explained the link between scientific research and world hunger. He remembered his own experience with too little food.

When the expected position with the Idaho National Forest Authority did not materialize, Borlaug decided to stay in school. He convinced Stakman to be both sponsor and adviser for his post-graduate studies. Stakman, in turn, convinced Borlaug to shift his major from forestry to plant pathology (the study of plant disease). This graduate degree would give Borlaug a broader base of knowledge. In 1941 when Borlaug was finishing his doctoral dissertation, he was approached by one of Stakman's former students. The scientist offered Borlaug a research position with a large chemical manufacturer. Borlaug accepted the offer and went to work in November 1941.

Borlaug's new job was to study the natural resistance of plants to various plant diseases. He was not directly involved with the production of any agricultural chemicals. However, in 1943, technicians in the same laboratory did some work on DDT. They tested samples of the insecticide for effectiveness on various insects. Later, Borlaug found out that DDT was used to control malaria-carrying mosquitoes in the Pacific and typhus-carrying lice in Europe.

In the mid-1940s, near the end of World War II, cases of typhus were reported in Naples, Italy. A powder form of DDT was applied directly to people's bodies and clothing to kill the lice. DDT was credited with preventing an epidemic of typhus and saving thousands—perhaps millions—of lives. No aftereffects from this intense and extended exposure to pure DDT were ever reported.

The Move to Mexico

In 1944, Stakman was once again involved with Borlaug's career. As part of the Rockefeller survey team, Stakman was asked to nominate someone to work on the control of plant disease in Mexico. He nominated Borlaug. At the time, Borlaug's job at the chemical company was considered essential because his project served the war effort. However, Rockefeller influence was strong, and he was released from the war-related project. By midsummer, Borlaug was free to join the scientists in Mexico City.

At first, Borlaug's project assignment was the control of fungus diseases of wheat called rust. He focused on improving the native Mexican wheat. However, he was also responsible for training young Mexicans to utilize new agricultural methods. Between 1945 and 1960, Borlaug helped educate more than 700 Mexican men and women. Over half of these students were sent to the United States or other countries for advanced university studies.

Borlaug believed in demonstrating new methods to his students. He laboriously studied Spanish so he could speak directly with the Mexican farmers. Gradually, the young Mexican students saw that success required mutual respect between the scientists and the hard-working farmers. This newfound respect frequently helped persuade farmers to adopt the proposed farming techniques.

After being on the job for only a few months, Borlaug took on new responsibilities for research on wheat breeding. He conducted the wheat-breeding studies in three main phases. During the first phase, he experimented with crossbreeding the 100 varieties of native Mexican wheat that George Harrar had planted in 1943. Borlaug hoped that the varieties that were resistant to rust fungus could be crossed with varieties that gave high yields. Unfortunately, none of the native Mexican wheats were particularly rust resistant or productive.

These findings led Borlaug to the second phase of his research. He obtained several samples of disease-resistant seeds from the African country of Kenya. Crosses between the African wheat and the best local varieties provided a breakthrough for Borlaug. His hybrid was superior to the parent plants. Borlaug encouraged the Mexican government to distribute the hybrid seeds throughout the country.

Borlaug now addressed the problem of mediocre productivity. This difficulty was due, in part, to the poor condition of Mexican soil. When nitrogen fertilizer was added to the soil, the wheat plants grew healthier and taller. However, the tall stems were fragile. Heavy wind or rain caused the stalks to bend and then the heads of grain lodged in the mud. Fertilizer alone could not correct insufficient productivity.

These conditions led to a third phase of wheat development. Borlaug obtained new seeds from a friend at Washington State University. The seeds were the product of a cross between a Japanese dwarf variety and three disease-resistant U.S. varieties. Borlaug used this semidwarf wheat for further crosses with his hybrid African-Mexican wheat. The outcome was a long series of new, improved wheat varieties that were disease resistant, highly productive, and responsive to fertilizer. In addition, the most recent of the varieties was highly adaptable to variations in climate and growing conditions.

Borlaug and his associates fought a long battle to achieve the cultivation of good Mexican wheat. Mexicans from all levels of society expressed their resistance to any change in traditional farming practices. High officials in the Mexican government resented outsiders who were attempting to improve productivity. Peasant farmers feared that any change in farm technology would bring about a greater risk of failure. Even within the Rockefeller Foundation, any drastic change from tradition was treated with anxiety.

In 1946, Borlaug thought of a new practice to improve the wheat-breeding project in Mexico. He suggested that the Rockefeller group fund two wheat crops each year rather than one. One crop would be cultivated near Mexico City and the other in Sonora. This change would take advantage of the difference in climate between Mexico City—in central Mexico—and Sonora—the wheat-growing region in northwestern

Norman Borlaug was a hunger fighter who worked closely with the farmers of Mexico. (Courtesy of the Agricultural Research Service, U.S. Department of Agriculture)

Mexico. Two plantings a year could reduce the time needed to develop, test, and harvest the seed for new varieties of wheat.

Permission was refused when Borlaug presented this idea to the director of the Rockefeller program. Borlaug was distraught and decided to resign and return to the United States. Fortunately, Professor Stakman happened to be present when the conflict became heated. Both sides had valid arguments. Borlaug's plan required periodic travel between Mexico City and the Sonora region. This was costly and dangerous because the journey involved about 2,000 miles (3,200 km) of bad roads and rugged terrain. The director also pointed out that wheat production in the Sonora area was high by Mexican standards and did not warrant additional study.

Borlaug argued that great quantities of time and money would be saved by doubling the yearly harvest of experimental wheat. In addition, he pointed out that the Sonora region would be gravely affected if a fungus epidemic should break out in the future. The densely cultivated fields of Sonora would be spared if they were planted with fungus-resistant seed.

Finally, Stakman negotiated a truce. Borlaug was allowed to plant his hybrid wheat in Sonora for one growing season. After he proved that the new plan saved time and money, his two-crop method was adopted and became a routine part of the official Rockefeller program.

As early as 1950, it was clear that the African hybrid wheats were truly resistant to fungus infections. The Mexican government announced the good news. Other countries in Latin America requested the seeds and asked Borlaug to train their students in the new techniques of plant breeding.

By 1956, Mexico produced more wheat than was needed for domestic use and began to export the surplus. However, Borlaug was not satisfied and began another phase of cross-breeding semidwarf varieties. The following year, new students began arriving from countries such as Afghanistan, Cyprus, Egypt, Jordan, Libya, Turkey, and Saudi Arabia, as well as 10 countries in Latin America. Semidwarf seed samples were sent to India and Pakistan in 1962, and Borlaug was invited to visit those countries the next year.

Unfortunately, the first Indian crop of semidwarf wheat seemed to be a failure. Tradition specified that only a small amount of nitrogen fertilizer could be used on wheat crops. The plants in the demonstration plots were weak and stunted. Borlaug was disappointed. He was unaware that some of his Mexican-trained, Indian technicians had disregarded tradition. In a small, hidden garden plot, the young workers had applied generous amounts of fertilizer after planting the semidwarf seeds. When the healthy plants had ripened, they were shown to top Indian officials. No one could deny that the bountiful wheat was the result of the correct use of fertilizer.

The government of India organized its own wheat production program in 1964. The following year, large-scale demonstrations were conducted on privately owned farms near the capital city of New Delhi. Comparison plots were planted at each farm. One plot was planted with local seed and cultivated in the traditional way. The other plot was planted with the Mexican-grown seeds of semidwarf wheat and given generous amounts of nitrogen fertilizer. The semidwarfs out-produced the native wheat by more than a 10-percent margin. During the years between 1966 and 1968, the Indian government distributed the new wheat seed throughout the country. By 1979, India had more than doubled its wheat production.

Meanwhile, Pakistan had also greatly increased production of the crop. The green revolution in Pakistan owed even more to the young scientists who had gone to Mexico for training. From 1963 to 1968, 30 of these young people directed the Pakistani program during the difficult early years.

A similar pattern emerged in North Africa and the Mideast. Turkey, in particular, mounted an effective program based on Borlaug's ideas. Later, good results were obtained in Brazil, Argentina, and the People's Republic of China. The dramatic success of Borlaug's programs led to his nomination for the Nobel Peace Prize. He was awarded the prize in 1970. The fame from this award led to his testimony before congressional committees about the use of agricultural chemicals, such as DDT. This testimony, in turn, led to his widely publicized confrontations with leaders of the environmental movement.

Apparently, Borlaug was worried that by condemning DDT the environmentalists would automatically condemn all agricultural chemicals. Therefore, he believed that he must defend the beneficial properties of DDT. He recalled how the chemical had successfully suppressed the malaria mosquito and the lice that carried typhus. Just as his grandfather had advocated, he remained true to his carefully considered convictions about DDT. The Nobel Prize that he had received in 1970 did little to ward off the opponents of his unpopular position. Critics of his stand wrote unflattering and sometimes untrue articles in newspapers and magazines.

The great irony of the whole episode centers on the fact that DDT is an insecticide. Wheat cultivation has little need for insecticides. Consequently, Borlaug had no connection to DDT in his work on improving the cultivation of wheat. The only agricultural chemical of interest to Borlaug was fertilizer.

Borlaug had been praised when he won the Nobel Peace Prize for his work in the wheat fields of Mexico, India, Pakistan, and elsewhere. A year or so later, many observers considered him a villain and a major propagandist for the chemical companies that make pesticides and herbicides. At the dawn of the environmental movement, he dared to criticize the ideas of Rachel Carson, the woman who had helped bring environmental issues into national politics. Borlaug's defense of DDT and his opposition to Carson's ideas were never understood. Opinion leaders in the United States harshly criticized the famed scientist for his views.

8

Ecological Science

Ecology is the study of the interdependence of plants and animals living together in communities. Ecologists emphasize that such communities require a hospitable environment to prosper and survive. They also state that there must be a continuing interaction between the communities and their environment. These and similar concepts help provide the scientific foundation for the philosophies and political goals of the environmental movement.

In order to understand the interdependence between organisms, ecologists measure the relationships among the resident plants and animals as they compete for essentials such as space, water, sunlight, and minerals. They also note the manner in which the plants and animals extract life-giving resources from their environment and how they cope with danger.

The idea that communities develop and change over time is one of the core principles of ecology. John Bartram of Pennsylvania, a well-read American farmer and naturalist, was one of the first people to give serious thought to this idea. From 1760 to the 1780s, Bartram studied plant communities and saw how such communities adjusted and regained stability after changes in their environment. He and his son, William, explored the wild areas of Georgia, Florida, and the lower valley of the Mississippi River. They wrote vivid accounts of their explorations that are still being read today. Modern scholars contend that his writings reflect what is now called a "systems approach" because of his emphasis on the recovery of balance after stress. Indeed, his works appeared to include the idea of an ecosystem but without the use of modern terminology.

Unfortunately, contemporary naturalists failed to develop Bartram's ideas. However, during the early 1800s, Alexander von Humboldt's work in South America contributed new thoughts on the methods and direction of nature studies. Humboldt's writings include the effects of environmental change on various plant and animal species. One of his studies recorded the types of plants found at various altitudes on a mountain. Humboldt discovered that certain plants prospered at a specific height while others failed. Although he is not often credited as one of the founders of ecology, his work can be seen as an early example in the development of that science.

Nature at the Shoreline and Below

The first pioneers who *were* recognized as founding the separate discipline of ecology were Earth scientists such as the Swiss geologist, F. A. Forel. Forel was interested in the variety of plants and

John Bartram and his son traveled widely along the Atlantic coast examining plant communities in many different environments. He wrote about his observations in ecological terms before the formal discipline was founded. (Courtesy of Historic Bartram's Gardens, Philadelphia, PA)

animals that live in freshwater lakes. His work, carried out in the late 1800s, followed the same research methods as those used by Humboldt. However, in place of the mountain slopes of South America, Forel investigated the submerged slopes of Lake Geneva in Switzerland.

Communities

The work of the pioneering geologists was followed in a few years by the innovative work of botanists. Until the early 20th century, botanists were busy locating, identifying, and classifying plants during their journeys around the world. About 1910, however, researchers such as Edward A. Birge, of the Wisconsin Geological and Natural History Survey, and Henry C. Cowles, of the University of Chicago, began to look at plants in communities rather than as individual representatives of particular species. These scholars encouraged students to examine the variety of species living together at a given site and to note the relative abundance of each species. For example, they questioned why a certain variety of bluegrass was dominant in a particular prairie setting while nettles and other broad-leafed species outnumbered all other plants in a slightly different—but nearby—location.

Areas that provide conditions that sustain a particular pattern of plants and animals are known as habitats. Indeed, a saltwater marsh habitat in the temperate zone is likely to be populated by plants and animals that closely resemble those found in saltwater marshes in other temperate areas.

The Biosphere

In the late 1920s, the Russian geologist Vladimir I. Vernadsky first discussed the idea of a biosphere. This concept states that a biosphere is composed of all the Earth's habitats—from rocky crags high in the mountains to the cold depths of the oceans—extending the scope of interdependence to a world-wide scale. For example, global processes that occur in the biosphere include the evaporation of water from the oceans, the subsequent formation of rain clouds, the release of rain over the continents, the river drainage back into the oceans, and to complete the cycle, the evaporation of water from the oceans.

More Interdependence

Around this same time, Charles S. Elton proposed the idea of the food chain. He saw plants as the essential providers of resources to communities of animals. Plants take minerals and water from the soil, carbon dioxide from the air, and—with sunlight as their source of energy—they produce sugars, starches, cellulose, proteins, and vitamins. The plants then serve as food for herbivores, or plant-eating animals. Herbivores then serve as prey for carnivores, or meat-eating animals. Humans generally occupy the top links of the food chain and are known as omnivores because they consume both plants and animals.

In recent years, the concept of the food chain has been revised. Scientists now recognize that the process is cyclical. Insects, worms, spiders, and microorganisms consume both the remains and the waste products of plants and animals. Thus, the chemicals, minerals, and water stored by plants and animals are recycled into the raw materials that nourish all species of plants.

Another revision of the food chain takes into account the fact that predators are likely to have more than one kind of prey and that a given type of prey may serve as food for many different kinds of predators. Thus, there is a whole range of predator-prey interrelations within a community. The complicated linkages are better seen as a web rather than a chain.

Analyzing Habitats

A different conceptual refinement was the development of the idea of the "niche." A *niche* is that portion of a habitat well suited to a particular species. The term reflects the suitability of a specific set of environmental conditions that provide for the well-being of a particular organism. George Evelyn Hutchinson was the first scientist to express the idea that a set of many factors—such as the amounts of various minerals in the soil, soil texture and drainage, and moisture level and temperature—must be combined to define the attributes of a particular niche for a particular plant. Different factors would be crucial for animals, but the same principle would apply.

Hutchinson's life spanned almost a century. He was born in 1903 in Cambridge, England, and died near the end of the 20th century. As a child, he showed an interest in the world of nature, and at the age of 15, he published his first article. His observations were concerned with the swimming ability of a grasshopper that he had observed in a local pond. It struck him as peculiar that a relatively large insect could swim with such ease. He recorded his ideas and submitted the note to a nature magazine. Hutchinson published many other research papers during his long life.

Early in his college studies, he was interested in all fields of science, but after graduating, his focus narrowed to zoology. Hutchinson received his advanced degree from Cambridge University and in 1925 was given a teaching position at a college in the Union of South Africa. This appointment did not meet with success, and he was dropped from the payroll after a year. However, the misfortune proved to be a blessing because Dr. Hutchinson was able to accept a research fellowship at Yale University in the United States. During the trip from South Africa to the United States (with a major stopover in Italy), Hutchinson immersed himself in some newly acquired books on life in freshwater environments—a science known as limnology. From then on, he regarded himself as a limnologist and made a career of limnological studies, using a pond near New Haven, Connecticut, as his main research site. He specialized in marking out the pathways along which nutrients were distributed in a freshwater environment. For example, he did a project that traced the flow of the mineral phosphorus from a single entry point to all locations in the pond and into the bodies of the resident plants and animals. Such research gave him

G. Evelyn Hutchinson, at about 60 years of age, as a Yale professor. He is holding a young potto, an unusual animal found in Central Africa. Pottos are related to monkeys but are more primitive. They are easy-going when young, but the adults, about the size of a house cat, turn mean. (Courtesy of the American Society of Limnology and Oceanography)

a prominent place in the difficult field of biogeochemistry—a field that combines research in biology, geology, and chemistry. During his many years of research and teaching, Hutchinson inspired students and colleagues by his vision of the connectedness of all the occupants of the natural world. Ironically, in 1977, the South African university that had cancelled his teaching contract named a graduate zoology laboratory in his honor. It is called the G. Evelyn Hutchinson Research Laboratory.

Based on Hutchinson's research, the concept of niche includes the consideration of all the elements in a habitat that benefit a particular species. Although a collection of different species occupies a habitat, each separate species in the community has its own niche.

More recently, ecologists are focused on the idea that dramatic changes in habitats can upset the combination of factors that result in a good niche for a given species. Such changes can lead to the endangerment of a species or even its extinction. Even small changes in a habitat can put pressure on the livelihood of a particular animal. For example, squirrels will eat almost any kind of seeds or nuts, but they seem to prefer acorns. Indeed, the squirrel population is likely to prosper when acorns are abundant. Conversely, they will suffer some privation if acorn supplies are in short supply and they must expend more energy than usual to find their favorite food. The energy used for finding food can reduce the energy needed for reproductive activities and cause a decline in the squirrel population in that habitat.

Only a decade after Hutchinson's arrival at Yale, another biologist, A. J. Tansley, made a formal restatement of Bartram's original conception of the ecosystem. Tansley saw that many conditions are necessary to maintain a natural state of balance. If one ingredient in the environment is in short supply—or is overabundant—natural forces tend to bring about an adjustment so that the community as a whole can survive. For example, research has shown that when forage is in short supply, female elks become less fertile. The population of elk then declines so that the herd can remain well fed.

Once a natural community is perceived as a web of interactions, such a community can be seen as an ecosystem in its own right. Such a system can grow and evolve over time, or it can deteriorate and be repopulated by other varieties of organisms. Tansley's concept of the rise and fall of certain species and the give and take among organisms—all in a framework of balanced stability—led him to revitalize his view of the ecosystem. The concept now means that communities of plants and animals can self-regulate and adjust to reasonable changes in conditions to survive.

A few years after Tansley's redefinition, ecologists were involved in a major disagreement. The argument stemmed from alternative ideas about the processes that take place in a profoundly disrupted ecosystem—such as one subjected to a natural disaster. The focus for the theoreticians was the concept of succession. Succession is the process whereby plant and animal species come into and occupy an area in which the original occupants have been removed or killed.

The dispute centered on the sequence of reoccupation. The scientists argued about which species would be the first to arrive after the disaster. One group of ecologists believed that natural restoration after a forest fire followed a very precise script. Such a script was supposedly decided by the specific attributes of plants. Plants that produced many seeds and grew rapidly would be the first to inhabit the land made vacant by the fire. Hardier plants that grew more slowly would follow. Another group believed that the processes of succession tend to be haphazard. They concentrated their attention on the species that prospered in the areas adjacent to the disaster and believed that the neighboring species could easily invade the vacant spaces. Recent observations such as those that followed the eruption of Mount St. Helens suggest that both positions have some merit but that the actual succession can be quite surprising. For example, the first organisms to invade the devastated areas were not plants of any kind. The invaders were thousands of infant spiders that were blown into the area riding on their spider silk parachutes. Not many of the baby spiders survived, but their remains provided an organic bed where airborne seeds were able to take root. Also, the spiders were attractive prey to large beetles that apparently survived the ash fall by burrowing into the dirt. Waste products of the beetles also contributed to the welcome afforded to the seeds.

Other evidence related to the succession controversy came from a comparative examination of the aftermath of forest fires. The sites of forest fires show various patterns of recovery. If the fire area contains many fallen trees and thick, dry underbrush, the fire will be hotter and more sustained. In such an event, even the strongest mature trees will be destroyed. The total destruction of a forest makes for an orderly but rather slow recovery. Scientists have observed an interesting aftermath of such circumstances. Plants known as fireweeds are likely to be the first vegetation to appear in the ashes of a

Mount St. Helens during its nine-hour eruption in 1980 (Courtesy of the U.S. Geological Survey)

forest fire. Their arrival is a welcome sign of the recovery of a devastated area.

At times, a forest or wooded area is purposefully set on fire. This activity is known as a controlled burn and is used when the chosen area has an overabundance of inflammable litter on the forest floor or many dried out shrubs and bushes. In a runaway fire, this litter could lead to an extremely hot burn. The Forest Rangers and firefighters monitor the controlled fire carefully and keep it small. Consequently, this type of forest fire is not as severe as a natural disaster and plants from adjacent areas have a good chance of colonizing. Pine trees whose seed cones open when slightly scorched will also have an advantage for repopulating the burnt areas.

Ironically, disasters such as fires and forest blow-downs—the destruction of forests that result when a very high wind blows down all the trees—provide prime opportunities for ecological research. Ecologists are generally reluctant to disrupt natural processes, and the intentional manipulation of an ecosystem for experimental research is avoided. Passive observation—studying a situation without interfering with it—is possible after a forest fire or a forest blow-down and is an ecologist's typical research method.

Differences in methodology limit the collaboration between ecologists and agricultural scientists. Ecologists study the interaction among plants and animals in the natural world and the dependence of all living things on environmental conditions. Agricultural scientists study the possibilities of improving plant species and crop harvests by employing new technologies and methods. Because farms are artificial environments, ecologists have rarely included such ecological systems in their research priorities. There are signs, however, that the situation is changing.

The Environmentalists Begin to Organize

9

As early as the summer of 1891, scientists and the public became interested in the insecticides sprayed by farmers on their fruits and vegetables. A green crystalline residue was found on grapes in a fruit store in New York City. The crystalline material was thought to be arsenic from Paris green, and a great fuss was raised by the New York newspapers. The New York Board of Health seized 250 crates of grapes at the wholesale fruit market. There was much embarrassment when the green material proved to be a simple mixture of copper and calcium salts used to control plant fungus. Scientists pointed out that an adult would need to consume 300 pounds (135 kg) of these grapes each day in order to receive a harmful dose of the copper/calcium mixture.

Eighteen years later, in Boston, Massachusetts, another health threat was uncovered. The city Board of Health found lead arsenate on pears that had been shipped from California. The fruit was immediately destroyed. The resulting uproar led to a general practice of checking all fruit shipped from the West Coast. However, a conflict of interests within the U.S. Department of Agriculture caused serious legal problems when tainted fruit was confiscated and destroyed. One section of the USDA instructed western farmers to spray insecticides on their fruit trees, while another part seized such fruit for violation of the pure food laws. In a landmark court case, the public health interests won, and suspect fruit continued to be destroyed.

The parties involved with this confusing state of affairs agreed that the conflict within the USDA could not continue. At first, officials advocated an informal agreement between the two sections of the department. The agreement stated that government agents must keep the contaminated fruit off the market until the owners had paid a small fine and washed the insecticide off the fruit. The public was not informed of these casual arrangements.

Meanwhile, foreign governments began to protest about the chemical residue found on imported produce. Fruit shipped to English distributors was often heavily contaminated. Officers of the British government threatened to ban imports of fruit and vegetables from the United States. The U.S. Department of Agriculture and the growers reacted vigorously. They took steps to ensure that all export shipments of fruit and vegetables were free from chemical residue. American consumers were unaware that produce sold in the United States was less carefully regulated than produce shipped abroad. If that condition had been known, the USDA would have experienced further embarrassments.

Although growers carefully followed the order to wash all produce to be sent overseas, some growers objected to washing fruits and vegetables to be sold in the United States. Growers began taking the USDA to court because they believed they were being unfairly treated by the government. They pointed out that they had followed the USDA's instructions on the use of insecticides. The growers were furious that they were being punished for complying with the law. The ensuing publicity threatened to give the Department of Agriculture a bad name.

In 1934, President Franklin Roosevelt appointed a special panel of experts to study the problem. This group, the Hunt Commission, recommended that only a very small amount of insecticide residue could be allowed to remain on fruit. Their suggestion was midway between the tight restrictions on exported fruit and the less demanding limits desired by American farmers and the Department of Agriculture. The secretary of agriculture, D. F. Houston, immediately suppressed the Hunt Commission report. Houston himself set the legal limits of insecticide residue. He authorized levels that could be easily achieved by fruit growers. Whether such levels were in the best interests of consumers has never been decided.

These early problems with insecticides indicate that a conflict existed long before the crisis in the 1960s. In 1962, the publication of Rachel Carson's book *Silent Spring*—a milestone in the environmental movement—documented an accumulation of knowledge about spray residues. This information justified the publication of a special, independent scientific periodical on chemicals used on food crops. It was originally entitled *Residue Reviews*. The journal, which continues to be published under the title *Review of Environmental Contamination and Toxicology*, gives evidence that the chemistry and biology of insecticides are now well-established scientific specialties.

The Nature of DDT

In 1874, Othmar Ziedler, a German chemist, developed a new chemical compound. DDT, as it is now known, is a carbon-based molecule with some chlorine atoms attached. Such chlorinated organic chemicals are often toxic. However, for many years no one knew (or cared) whether the new compound was toxic or safe. In fact, no one knew whether DDT was useful for any purpose. In the early 1900s, a Swiss chemical company purchased a group of patents from Ziedler's estate that included the manufacturing rights for DDT. Years later, one of the Swiss company's chemists discovered that DDT was a very effective insecticide.

In 1942, at the height of World War II, officials of the Swiss firm smuggled some of the material out of German-occupied Europe. The sample was brought to the United States and analyzed at the U.S. Department of Agriculture. The material was tested for its effectiveness against insects and as a possible health hazard to humans. It scored well on all the tests for short-term effects. USDA scientists found no health problems after volunteers were exposed to large amounts of DDT—either mixed with kerosene or in a dry powdered form. Later tests continued to indicate that there were no short-term medical effects on humans. No tests for long-term effects were administered at that time.

During the war years, when the need for insecticides was grave, the long-term effects of DDT were not considered relevant. Therefore, after the brief testing period, DDT was used extensively in the South Pacific to control mosquitoes. The frequency of malaria among the Allied soldiers was greatly reduced. The insecticide was also used to suppress lice among the civilian populations of displaced persons in Europe. A predicted outbreak of typhus was avoided. DDT had such a dramatic effect on the control of these two major diseases that Paul Muller, the Swiss chemist who first

Magnified head of an anopheles mosquito. The long, specially adapted mouthparts of the female are arranged to penetrate the skin and provide a channel for ingesting the blood of the victim. (Courtesy of the Office of Communication, U.S. Department of Agriculture)

discovered its usefulness, was awarded a Nobel Prize for medicine in 1948.

During World War II and for some years after, DDT was regarded as a godsend. Indeed, no one saw any disadvantages to the insecticide, and public opinion generally favored its use. Consequently, DDT was seen by many local government officials as a simple way to solve minor problems such as urban mosquito infestations.

Rachel Carson Enters the Fray

In the late 1950s, Rachel Carson, a distinguished science writer, learned about a disturbing event. A friend of Carson's had observed the appalling consequences of a pesticide program used in a Long Island, New York, community. The chemical DDT had been sprayed from aircraft on an infestation of mosquitoes. Carson was told that many birds died from the spray, but that the swarms of mosquitoes seemed to be as thick as ever. Officials had responded to the problem by increasing the concentration of DDT. This solution did not make sense to Carson or her friend. Carson decided to attack the problem.

Carson's public reaction to the use of the insecticide caused a major reversal in public opinion about DDT and other agricultural chemicals. Soon people from all levels of society were against the use of these substances. Since DDT had been approved and appreciated by most people in the United States, only someone of Rachel Carson's stature and ability could have brought about this rapid change.

Rachel Carson was born in 1907 in the small town of Springdale, Pennsylvania. In 1925, she received a modest scholarship and began her studies at what is now Chatham College in Pittsburgh, about 15 miles (24 km) from her home.

Initially, she chose English literature as her major field. In her second year, influenced by an inspiring teacher, she decided to major in biology. The two usually separate interests—literature and science—would direct the rest of her life. She became one of America's most celebrated nature writers.

When she graduated from college in the summer of 1929, she was hired as a research assistant at the Marine Biological Laboratory in Woods Hole, Massachusetts. It was the first time that Carson had seen the ocean, and she was greatly impressed. That fall she began her graduate studies at Johns Hopkins University in Baltimore, Maryland. Three years later, Rachel Carson received a master's degree in oceanography.

By the time she graduated, the Great Depression had ruined the economy, and she was worried about finding a job. Carson had some contact with the U.S. Bureau of Fisheries, then part of the Commerce Department. Luckily, they needed a qualified scientist who could write for nonscientific audiences. Her first project was a series of seven-minute radio scripts called "Romance Under the Waters." Later the fifty-two scripts were reorganized and submitted to the *Atlantic Monthly,* a very influential magazine. She was paid $75 when the scripts were published.

This success encouraged Carson to do more freelance writing in her spare time. Her first book, *Under the Sea Wind,* came out in November 1941. Reviewers praised the

Rachel Carson blended modern science with the sensitivities of a traditional naturalist. (Courtesy of the Lear/Carson Collection at Connecticut College)

book, and a book club sold it through direct mail order. The timing was poor, however, because the United States was soon involved in World War II. The first printing of Carson's book sold fewer than 2,000 copies.

During the war years Carson worked for the Fish and Wildlife Service of the U.S. Interior Department. She wrote pamphlets to encourage people to eat fish instead of red meat. Her writings also supported the value of newly created wildlife refuges such as those proposed by the naturalist Aldo Leopold. In her spare time she wrote nature articles for magazines and newspapers but did not begin a new book.

After the war Rachel Carson was anxious to resume her career as an author of nature books. She wanted to write the epic story of the Earth's oceans and how they became the cradles of life. This effort led to *The Sea Around Us.* The book was published in 1951 and remained a best-seller for almost two years. Its success led to the reissuance of *Under the Sea Wind,* which also became a best-seller. The financial returns from these books allowed Carson to resign from government service in 1952.

Her next book, *The Edge of the Sea,* was published in 1955. Reviewers said it was the first book that communicated the main ideas of ecology to a general audience. Carson's work documents the interdependence of the plants and animals that occupy a defined location. The book tells about the intricate web of life in what is now called an ecosystem. *The Edge of the Sea* also became a best-seller. Carson was a national celebrity.

Her next project began after hearing her friend's disturbing account of spraying an area on Long Island with DDT. In some ways this new project was a logical outgrowth of her preceding works on ecology. However, in some other ways it was a radical shift in point of view. The previous books were about natural processes. The new work, *Silent Spring,* focused on the improper use of synthetic pesticides.

At first Rachel Carson had hoped to write a magazine article on the subject. She believed that this would be the quickest way to generate public interest in the problems of DDT. However, her article was too controversial to be accepted for publication. Finally, the editors of *The New Yorker* made a commitment. The magazine, which had printed long excerpts from her previous books, agreed to use her article as a book preview. Consequently, Rachel Carson had to agree to research and write another book rather than just a single, short article.

Carson knew that her work would be attacked and that she would provoke powerful enemies in the chemical industry. This certainty caused her to invest an unusual amount of time in conducting research on DDT and other agricultural chemicals. The book was published in September 1962. The project had taken almost seven years to complete.

Silent Spring stated that molecules of DDT and related pesticides did not break down into harmless materials in the soil or water or after being ingested by animals. Because of this, the chemicals accumulated and became more and more concentrated in the food chain. Insects with small quantities of DDT in their bodies were eaten by birds and fish. Since birds and fish could not excrete all the poison, it built up in their bodies. When larger predators, such as humans, consumed the animals, the same process occurred—the poison accumulated. Rachel Carson believed that these facts provided a strong reason to control the use of such pesticides.

As expected, Carson encountered attacks from companies that manufactured pesticides. Many farmers and others in the business of agriculture were convinced that a ban on DDT would harm their prosperity. The dispute resounded in newspapers, in the courts, in government regulatory agencies, and in the Congress of the United States. In a weak attempt to stifle the controversy, congressional leaders organized committees to hold public hearings. Norman Borlaug gave testimony in the later rounds of hearings. Although Borlaug and Carson never met face-to-face, reporters made it appear that a direct confrontation had occurred between these two famous people. Newspaper and TV commentators played up the differences in their views. Indeed, the news media circulated claims and counterclaims from both sides of the DDT controversy.

For five years after the publication of Carson's book, those in the business of agriculture continued to employ DDT. Although citizen's groups lobbied Congress in an attempt to suppress the use of the insecticide by legal means, their efforts were fruitless. The only real progress was a reduction in government-sanctioned aerial spraying of cotton and other crops. Indeed, to make things worse, some farmers resorted to using more toxic insecticides when they were pressured to stop using DDT. Environmentalists seemed to be losing ground, even though public opinion was firmly on their side.

Both the court system and government agencies—such as the Public Health Service—had been slow to act on the pesticide threat. In 1966, a group of lawyers and scientists sought to hasten the limitation of DDT use. They took their test case to the Long Island jurisdiction that had prompted Rachel Carson to write *Silent Spring*. The judge acknowledged that DDT was probably harmful to the environment and restricted the use of DDT in the county for one year. The judge would not go further because he believed that existing laws were unclear. He stated that new laws were needed if the use of DDT was to be permanently restricted. The judge concluded that pesticide regulation was a problem for the state legislature, where laws were written, but not for a court, where a judge interprets the laws.

After the court failed to provide a permanent solution to the problem, many citizens were angered. Local government officials—once supporters of DDT—sided with the environmentalists. They recognized that their political careers were at risk if they did not respond to public opinion.

The next year, 1967, these same citizens resolved to monitor all proposed insect spraying and other uses of DDT. The group established the Environmental Defense Fund (EDF). The bylaws of the EDF allowed the organization to bring damage suits against any local government agencies that employed DDT as an insect control.

In the meantime, field studies in biology provided more disturbing evidence against DDT. Researchers found a sudden decline in the population of duck hawks in the North American flyways. They determined that the adult birds were not being killed by DDT, but that the chemical was probably interfering with the birds' capabilities to mate and reproduce. This news raised new questions about the possible adverse effects of DDT on the reproductive functions of humans. Many who had been indifferent to the fate of songbirds quickly became personally interested in the environmental movement.

Later in 1967, the lawyers and scientists who had been involved with the unresolved court case on Long Island decided to change their tactics. Some powerful Michigan politicians had become committed to the movement, and EDF lawyers sought to try their test cases in that state. The lawyers argued that the citizens of a given jurisdiction had a legal "right" to a clean environment. Although this was a weak legal argument and the EDF was not speaking for a specific group of citizens, the environmentalists had some success. Some small Michigan jurisdictions settled their suits out of court because they did not want to spend money on defending DDT. However, when one of the cases went on trial, EDF lawyers faced an unexpected problem. The judge was reluctant to concede that the EDF had the legal standing to represent citizens' groups who had not specifically hired the EDF team as their lawyers. The issue was not decided because the suit was settled out of court.

Following this suit, the EDF brought another case before a Michigan court. The lawyers sought to prevent the pollution of Lake Michigan by runoff from DDT and dieldrin (another persistent insecticide). The EDF again achieved only a limited ruling. A temporary restraining order prevented that specific instance of DDT use. However, the goal of the EDF was not achieved. They wanted to establish a legal precedent that would guide all decisions about the use of DDT. They wanted a judgment that could be used again and again in cases around the country.

In 1968, the EDF moved their activities to Wisconsin and brought a lawsuit against the city of Milwaukee. The lawyers sued the city to prevent the use of DDT in controlling Dutch elm disease. The case did not come to court because the city gave in during pretrial hearings. Again, the EDF people were disappointed with the outcome. They had hoped to win their suit and establish a vital legal precedent.

The attorney from the city of Milwaukee was sympathetic to the EDF cause and suggested that they try a different approach. He told them of a Wisconsin law that allows citizens to request a statement of the reasons for enforcing any government regulation. This procedure was established to make sure that all administrative regulations were understandable by the public. The law proved to be the ideal means to make a public case against DDT.

The members of a local environmental group questioned the Wisconsin Department of Natural Resources as to whether the state had a regulation in which DDT was defined as a water pollutant. If no such definition existed, the department was required to hold a hearing to answer the question concerning the nature of DDT.

The EDF planned 10 days of presentations in front of a panel formed by the Natural Resources Department. Both experts and public witnesses were to testify. All the witnesses—such as Gaylord Nelson, then U.S. senator from Wisconsin—spoke about the environmental dangers of DDT and its possible threat to human health. The EDF also argued that there existed safer ways to control insects than by spraying DDT or other toxic insecticides.

The hearing was scheduled for the first weeks in December 1968. The DDT defense team—from the National Agricultural Chemicals Association (NACA)—was taken by surprise when the EDF revealed its strategy. The professional Washington lobbyists had considered the EDF as a group of sentimental bird lovers and had not taken the case seriously. Therefore, NACA lawyers were ill prepared to argue against the mass of testimony prepared by the EDF. When representatives of the NACA were cross-examined by the EDF attorneys, it became clear that the NACA had a very superficial knowledge of the chemical and biological nature of DDT. The EDF won its first clear-cut case against the use of DDT. The situation was widely and colorfully reported in the newspapers.

The EDF also announced that the U.S. Department of Agriculture gained its information on chemical safety from research conducted by the chemical manufacturers themselves. This announcement caused great amusement with comments about putting foxes to work guarding the chicken coop.

Earlier in the year, the General Accounting Office—a watchdog agency responsible to the U.S. Congress—had reported another problem with the USDA. They found serious deficiencies in the way that the USDA enforced the Federal Insecticide, Fungicide, and Rodenticide Act. The USDA's enforcement problem and the questionable studies on chemical safety forced many government officials to consider transferring the USDA's enforcement responsibilities to another government agency. After legislators learned of the concerns, Congress passed the National Environmental Policy Act of 1969. This law required an environmental impact assessment for every major developmental project undertaken by a federal agency. Although the act did not address the conflict of interest within the USDA, the regulations helped legislators and the general public understand what officials within the USDA were doing.

With the publicity generated by the case in Wisconsin and the follow-up actions of the EDF, the idea of environmental protection became more popular. The discussions in Congress and among high officials of the federal government were closely watched by the citizens' groups and the mass media.

During the late 1960s, however, state and local governments showed more interest in environmentalist causes than did the national government. Cities and towns in New York and Michigan wrote prohibitions on the use of pesticides. Then state legislatures in Wisconsin, Michigan, Washington, Maryland, Vermont, and California passed such laws. After actions at the state and local levels, politicians at the national level began to pay more attention to protecting the environment.

In the mid-1960s, organizations such as the Sierra Club and the National Wildlife Federation had fewer than 500,000 members. Their political power was small since the membership was scattered over the United States and included many divergent political views. Soon, however, membership dramatically increased. College students, who had demonstrated against the war in Vietnam and mobilized for civil rights, found a new cause to support. The environmental organizations began to sponsor special activities on many college campuses.

On April 22, 1970, millions of United States citizens—young and old—participated in the first Earth Day celebration. The Sierra Club and other environmental organizations had been the forces behind this and other political demonstrations. Politicians such as Senator Mike McCloskey of California endorsed the new movement. The politicians realized that thousands of their constituents had joined the cause and could easily vote against unsympathetic officeholders.

Soon more national politicians became involved in the environmental movement. In late 1970, President Richard Nixon established the Environmental Protection Agency (EPA) by presidential order. Two years later, William Ruckelshaus, who had been named as the director of the EPA, announced a general ban on the use of DDT.

Almost exactly 10 years had passed since the publication of *Silent Spring*. Unfortunately, the person who had initiated the environmental protection movement did not live to see the realization of her goal. Rachel Carson had died of cancer in April 1964.

10
Rice
The Green Revolution Continues

More than half the people in the world depend on rice as their principal food. Many people on the western rim of the Pacific Ocean consume rice every day of their lives. Indeed, they eat rice at every meal. The typical Asian eats more rice in a week than the average American consumes in a year. In a country like Thailand, where rice has been grown since prehistoric times, the cultivation and consumption of rice has both mythic and mystical properties. Rice is not just a food; it is important to the total culture.

John D. Rockefeller III made the study of Asian culture an important part of his life. After World War II, Rockefeller was particularly interested in the revitalization of Japan. He knew that the modernization of farming methods aided a country's economy. Millions of people had been helped when scientists brought about a revolution in corn production in the 1930s and wheat production in the 1950s. In the 1960s, the Rockefeller Foundation developed a plan to improve the productivity of rice in Japan and all of Asia.

America's political involvement in southeast Asia was a very delicate matter. Rockefeller was careful to avoid programs that conflicted with U.S. foreign policy or with projects undertaken by the United Nations or other multinational bodies. Moreover, the administrators of his family's foundation were aware that influential people from around the world were opposed to any American involvement in Asian economics. Some of these people benefited financially from the low wages paid to many Asian workers.

Although most nations and individuals believed that fighting hunger was above criticism, a few were upset by Rockefeller's involvement in this cause. They saw an outsider interfering with Asian affairs. In Rockefeller's view, a plentiful food supply led to political stability. This stability provided the opportunity for economic and cultural recovery in the postwar era. However, not every powerful person in Asia was in favor of this stability.

John D. Rockefeller III saw the way to meet increased needs for food in the Far East. (Courtesy of the Rockefeller Archive Center)

Aware of the forces against him, Rockefeller moved forward with his plan to improve the cultivation of rice. He decided to model the new venture on the Mexican program—the successful crossbreeding of wheat plants. He hoped to find a host country in Asia and duplicate his good relationship with the Mexican government. The project needed a home base and the cooperation of local government officials. Fortunately, officials of the Philippine governments offered space on the campus of a small agricultural college in Los Banos, about thirty-four miles (54.4 km) from the capital, Manila.

Rockefeller again chose George Harrar as coordinator of a project. In 1958, Harrar took an extended leave from his post near Mexico City in order to survey the rice-growing regions of Asia. His report to Dean Rusk, the Rockefeller Foundation president, confirmed the need to increase rice productivity. Harrar's report also stated that a variety of good quality rice plants were available for breeding experiments.

In the meantime, Ford Foundation officials were also looking for a new project to reduce hunger and disease. Soon the Ford Foundation, already active in Asia, and the Rockefeller Foundation formed a partnership to improve the supply of rice.

As soon as the two foundations agreed to work together, they invited representatives from the Philippines and other Asian governments to participate in the project. Work at the International Rice Research Institute at Los Banos was soon under way. Representatives from Japan, India, Thailand, Taiwan, and the Philippines were invited to sit on the governing board of this body.

The board members employed Raymond Chandler to be the program director. He hired a staff of senior scientists that soon numbered more than seventy people. Chandler, who had taught agricultural science at both Cornell University and the University of New Hampshire, made arrangements to bring in promising young students from many Asian nations. In addition, he directed the construction of new buildings at Los Banos. The Ford Foundation contributed over $7 million for that purpose. The resulting facilities were so well designed that Los Banos became a popular tourist attraction in the Philippines.

Getting Down to Work

In 1962, the crossbreeding of rice plants began at Los Banos. Reflecting the lessons provided by the wheat project, scientific attention was focused on the cultivation of short-stemmed rice plants. The breeders knew that semidwarf grain varieties respond well to nitrogen fertilizer. When well fertilized, each rice plant grows many short strong stems, and each stem produces a good head of grain. With fertilizer, tall varieties of rice produce gangly fragile stems. The tall stems soon bend or break, and their rice grains are lost in the mud.

The researchers at Los Banos took only four years to develop their first success. It was a semidwarf rice plant designated as IR-8. In 1966, seeds were distributed to farmers in the Philippines and other countries. These seeds produced plants that were highly productive and also disease resistant.

For better or worse, the International Rice Research Institute had a very active public relations department. These publicists made sure that their product received immediate

This semidwarf rice, having high-yield capabilities, was one of the first to be developed. The rice plants also matured quickly—permitting two crops in one year. (Courtesy of the Agricultural Research Service, U.S. Department of Agriculture)

media attention in the United States and Asia. The publicists—and then the press—described IR-8 as a miracle rice. Throughout Asia this glowing label raised hopes that could never be fully realized.

There are many problems associated with the production of rice. First, rice is cultivated under a great variety of growing conditions. Climates vary from subarctic in northern China to tropical in Sri Lanka. Soils, rainfall, and cultivation methods vary greatly. The differences in local growing conditions are so dramatic that rice grown in one valley will not prosper in the next. IR-8 did well in the fields around Los Banos, but did less well a few dozen miles away.

Another problem in rice culture arises from the wide variety of consumer preferences. People from Asian countries place great importance on the taste of rice. As might be expected, the taste of IR-8 did not appeal to all Asians. In addition, the consistency of cooked rice is significant. Some people like sticky rice, while others demand a rice in which the kernels remain separate. The appearance, size, and texture of the rice kernels are also important factors. White rice—rice kernels with the husks removed—is almost always preferred over brown rice—kernels with some residue of the husks untouched. Unfortunately, the kernels of IR-8 cracked when the rice was milled to remove all of the husk.

Asian farming practices also caused problems for the scientists at Los Banos. The IR-8 rice grew best when provided with plenty of water, nitrogen fertilizer, and insecticides. However, many Asian farmers could not afford to treat their rice paddies (the submerged fields where rice is grown) with the agricultural chemicals suggested by the experts. The experts and rice farmers were faced with still another problem. As with all grains that are bred for disease resistance, this characteristic gradually faded from IR-8. The predators soon adapted and regained their power to decrease the rice harvest.

In spite of these problems, IR-8 was a success. Rice production in the Philippines increased by 10 percent the first year after IR-8 seeds were cultivated. Other countries with good irrigation systems also benefited from the cultivation of IR-8.

The researchers at Los Banos knew that IR-8 would not fill every need and continued to develop new plants. Thus IR-20 came out in 1969, IR-36 was released in 1976, and IR-56 was ready a few years later. Each of these new varieties was a high yield, semidwarf rice plant with greater pest resistance than its predecessors.

A Mixed Success

These new rice varieties have been beneficial to the peoples of Asia. In most Asian nations rice production continues to increase and supply has kept pace with demand. Even the poorest farmers have profited from the new strains of rice. Moreover, city dwellers in such places as Bangkok, Thailand; Singapore; and Jakarta, Indonesia, also benefit because the price of rice has remained low.

From a technical perspective, the progress in rice production paralleled the progress in wheat production. Both programs included crossbreeding to achieve specific traits such as short stature and disease resistance. Fertilizers are of prime importance to the productivity of both grains. However, the use of insecticides has caused many problems with the cultivation of rice.

Indonesia developed a particularly bad situation. Pests that feed on rice plants were quick to develop an immunity to the prescribed insecticides. Within a few years, the swarms of pests were larger than before the use of the agricultural chemicals. Many Indonesian farmers feared that their crops were being sabotaged. The American and Indonesian technicians had to make some major adjustments. They looked to the farmers for guidance and began to try various forms of natural pest control. The scientists changed the planting schedule so the rice could be harvested before insects called brown hoppers were mature. Gradually, the situation improved with the use of additional changes in planting and harvesting practices and the use of natural pesticides. Once again, Indonesian farmers could benefit from the new, highly productive varieties of rice that had been developed in Los Banos.

The revolution in rice production continues to this day. Almost every nation in Asia now employs its own research and development projects to improve the cultivation of rice. In the meantime, American scientists remember the lessons they learned in Asia. After their problems in Indonesia, many were more inclined to consider natural methods of pest control. This and other agricultural strategies were carried back to the United States.

11

Environmental Reforms

During the late 1960s, the U.S. Congress began to receive more adverse information about DDT. In 1972, Congress amended the original Federal Insecticide, Fungicide and Rodenticide Act of 1910 (FIFRA). The amended law was called the Federal Environmental Pesticide Control Act. The original law, designed by the USDA, protected farmers from worthless or dangerous pesticides by forcing manufacturers to list the ingredients on the labels. The amended law protected *all* citizens. It authorized the Environmental Protection Agency (EPA) to regulate pesticides. The director of the EPA was empowered to ban or restrict the use of any insecticide that appeared to present a danger to nontargeted plants and animals.

Environmentalists were pleased when the regulation of pesticides was expanded and transferred from the USDA to the EPA. They had been concerned about a conflict of interests. The USDA represents agriculturalists who are in the business of producing plentiful food at affordable prices. Not surprisingly, many of these people favored unlimited pesticide use. When the USDA controlled the enforcement of pesticide regulation, this was a difficult situation. Many felt that the USDA would be sympathetic toward the viewpoint of the agriculturalists. Most groups agree that the EPA has avoided this conflict and represents the interests of all citizens. However, some environmentalists saw the need for more changes.

Special Interests

The proposed changes took many forms—some of which were not broadly popular. For example, some groups were interested in regulations aimed at human food consumption. They advocated that everyone should eat less meat. These groups regarded meat consumption as harmful to health and against the ethics of animal rights. Some considered eating meat to be an inefficient use of raw materials and stressed that nine pounds of corn are needed to produce one pound of beef.

A few environmentalist groups suggested more radical proposals. A small number wanted the government to give financial aid to people who sought to leave cities and relocate in the country. These people would be given small farms and encouraged to use environmentally sound methods of cultivation.

Enforcement Problems

Most environmentalists do not promote such novel ideas. They propose that the existing laws should be enforced. However, law enforcement suffers because the EPA and other agencies have more work than they can handle. In a typical year, more than 200 new pesticides are developed. Each pesticide might contain one or more ingredients from a list of over 600 toxic substances. Therefore, each new product should be carefully analyzed before it is approved by the EPA. This would be a mammoth undertaking.

The EPA does not have the money or human resources to analyze all new products. To a large degree, EPA approval of each pesticide still depends on the testing done by its manufacturer. Based on such information, the EPA approves many products for sale. Originally, if a product was found harmful after its approval, the EPA was required to purchase and destroy all the previously manufactured material. This provision of the law made the EPA reluctant to ban a product once it had been approved and placed on the market. In 1988, the requirement was changed, and the manufacturer was made responsible for the cost of destroying harmful materials.

These safeguards do not deter some environmentalists from wanting to ban all chemicals from agricultural use. To illustrate the need for this prohibition, they point out that agricultural chemicals are an important source of stream contamination. Chemical fertilizers and insecticides are washed off crops by rainwater. The polluted water flows into streams rather than into wastewater treatment plants and thus remains contaminated. Both humans and wildlife may suffer

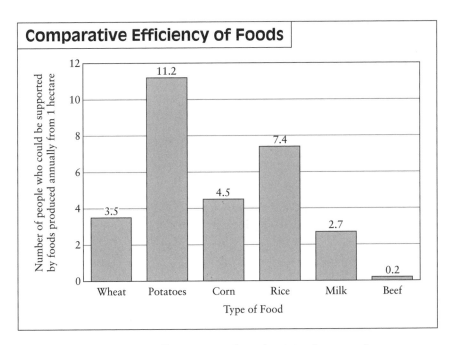

Comparative Efficiency of Foods

Number of people who could be supported by foods produced annually from 1 hectare

Wheat	Potatoes	Corn	Rice	Milk	Beef
3.5	11.2	4.5	7.4	2.7	0.2

Type of Food

Eating red meat is an inefficient use of productivity because the same amount of land used to grow cattle feed could produce enough wheat, potatoes, or corn to feed more people than the beef would feed.

Runoff Pollution

Fertilizer is being applied to a field.

Fertilizer runoff into nearby waterways increases algae growth that kills fish.

Agricultural drainage can include fertilizers, which increase growth of algae-depleting oxygen needed by fish in bodies of water, and pesticides, which can kill plants and animals.

from contact with the untreated water. Environmentalists believe that this type of pollution is unacceptable.

Commercial food producers do not agree that environmental problems such as stream pollution outweigh the benefits of agricultural chemicals. They claim that without insecticides many agricultural products would be destroyed or damaged by pests. Food producers argue that damaged fruits and vegetables do not appeal to consumers and will not sell. They maintain that pesticides must be used to avoid food shortages and the loss of profits.

Reconciliation

However, environmentalists and commercial food producers now agree on one important issue. They both believe that extensive or uncontrolled use of synthetic pesticides is futile. Both groups know that some pesticides can destroy other creatures while eliminating targeted pests. They also realize that pests adapt to pesticides.

When a pesticide is used, a few pests always survive. The survivors will be the parents of the next generation. Those offspring will inherit their parents resistance to the pesticide. Historically, farmers used increasing amounts of pesticide in an attempt to eliminate the resistant pests. This approach is flawed. Studies have shown that *no* amount of pesticide can control all the descendants of the originally targeted pests. In addition, too much pesticide can kill the pest's natural enemies.

Agricultural scientists are attempting to design programs of pest and weed control that will avoid the problem of pesticide resistance. They now use a wide variety of farming techniques and a minimum of synthetic chemicals. The approach—called integrated pest management, or IPM—employs natural products to fight pests.

Integrated Pest Management

Unlike some earlier efforts at pest control, IPM does not attempt to destroy every individual agricultural pest. Such a goal is impossible. The objective is to reduce pest populations to the point where their damage can be afforded by food producers.

Among the most appealing natural pest controls is the application of natural pesticides. Some plants such as chrysanthemums produce such chemicals. These natural chemicals can be extracted and concentrated for use as sprays or powders. They do not accumulate in the body like DDT, but quickly break down into harmless materials. Natural pesticides control certain pests, but do not harm other living things.

This technique is environmentally safe, but not, at present, commercially viable. In their search for natural insecticides, scientists have tested a large number of plants—as many as 15,000 varieties—before finding an effective product. In addition, a product might control only pests that attack minor crops. Rotenone from chrysanthemums is an example. It is not useful for treating major crops such as corn and wheat, so the developer's financial return will not justify his or her costs.

In some cases, pests can be controlled by trickery. Recent research has revealed that many insect species find their mates by smell. The attractive smell produced by the insect is generated by a chemical known as a pheromone. The pheromone—and therefore, the smell—is specific to each species. These chemicals can be identified and synthesized in the laboratory. The pest control manager uses small amounts of the chemical to lure the male insect to his death or lead him into a trap. Indeed, males of some species exhaust themselves by attempting to mate with objects that have been dosed with the right pheromones. Since pheromones are not toxic, this technique is a danger to the targeted pests but not to the environment.

Anther form of deception has been used to combat insects that attack soybeans. A small plot of land, perhaps 30 feet by 30 feet (10 m by 10m), is seeded a week before a large, adjacent field is planted. The plants on this small plot will reach maturity and attract the pests while the main crop is still immature. The swarm of pests in the small plot can be attacked and killed with a minimum amount of pesticide. When the main crops mature, they are relatively free from invaders.

A more advanced variation on this technique is to plant a small, untreated plot next to large, cultivated fields that have been treated with synthetic insecticides. The *untreated* plot provides a feeding ground for the hungry pests, and few of them attempt to eat the treated plants. Therefore, a relatively small number of insects ingest the insecticide. This tactic lessens the chances that the target pests will develop resistance to the agricultural chemicals.

Other means of control induce insects to enter traps and similar destructive situations by other types of attractants. An increasingly common device used in hospitals and industrial kitchens employs a special light source. The brightness of this light is particularly attractive to some flying insects, such as houseflies. The light source is mounted inside a wire cage. The wires carry an electrical charge. When the insect flies between the wires of the cage, an electric circuit is completed and the charge kills the insect.

Attractants are also derived from insects' normal food supplies. A small piece of rotten fruit is a good lure for houseflies. The fruit is placed in a box trap made of fine mesh. The opening is funnel shaped—easy to enter but difficult to escape.

One of the main lines of scientific research that is needed for the further advancement of integrated pest management is the production of more detailed descriptions of the life patterns of pests and those of their predators. For example, wheat fields in the western states suffer from infestations of aphids—some of which were accidentally brought here from Russia. University researchers have imported some of the aphid's natural enemies. One such enemy is a predatory wasp that lays its eggs in the aphid's body. However, the life cycles of the aphids and the wasps are not always in synchrony. Since the wasps are raised in an artificial environment, it is possible to time their maturity and release to coincide with the maximum vulnerability of the aphids. Several factors determine when the aphids can be attacked most effectively. These include the amount of rainfall and high and low temperatures in a 24-hour span. Researchers can make good predictions of vulnerability when they have the details of the life stages of the aphids.

Meanwhile, both the U.S. Department of Agriculture and the Environmental Protection Agency (EPA) have been working to spread the adoption of integrated pest management ideas and practices. For agriculturalists, moving forward has been in the hands of county extension agents who can talk directly to farmers and explain the advantages and the technical issues. Since the EPA does not have such a group of communicators, officials needed to find alternative communication channels. EPA officials found one answer in the public schools. Technical specialists at the EPA developed a model curriculum for teachers. This teaching plan can be scheduled as part of the normal science offerings of a school at various levels—but emphasis is given to students in kindergarten through sixth grade.

The outline of the teaching plan includes an explanation of the need to reduce the use of chemical pesticides and a description of the procedures for alternative methods of pest control. In the steps that follow, the focus narrows to specific pests. For example, roaches are likely pests in schools that contain lunchrooms. There are chemical products designed for roach elimination but alternatives exist. The use of chemicals can be reduced if certain sanitary practices, such as the prompt removal of food scraps and waste, are followed.

By reaching out to the schools, the EPA achieves several objectives at the same time. First, a bridge is built between the students and their families on one side and the farmers on the other. Second, with an emphasis on actions within the school building, the ideas and practices are easily transferred to the houses where the students live and to their neighborhoods. In sum, the school-directed programs pull together the rural and the urban families in a common effort to control pests and vermin with a minimum impact on the quality of the environment.

School projects were launched in the mid-1990s and continue to the present by means of grants-in-aid provided by the EPA's Environmental Stewardship Program. In 2003, projects were put underway to promote the introduction of integrated pest management ideas on a statewide level. Most such projects are carried out by people who do research on educational techniques and who are teamed with specialists in pest management. Current efforts include the adoption of instructional presentations on CD-ROM or DVD formats.

Repellents

Citronella, a natural product used to repel mosquitoes, illustrates how a certain odor can appeal to some pests and repulse others. Although mosquitoes dislike the smell of citronella, oriental fruit flies are strongly attracted to it. Therefore, the same natural product can act as a lure or repellent, depending on the targeted pest.

Synthetic products can work in a similar manner. Kerosene— a repellent to some pests—attracts the Mediterranean fruit fly and can be used as a lure.

Other new techniques can provide plants with the ability to be distasteful rather than toxic to predators. For centuries farmers have known that some plants repulse unwelcome feeders. Scientists can now duplicate this natural capacity by extracting the active ingredient from the plants. However, obtaining sufficient quantities of the repellent appears to be just as difficult and expensive as extracting natural toxins, such as rotenone.

Agricultural chemists might also obtain natural repellents by synthesizing the active ingredients. At present, this method does not seem practical. Duplicating and mass producing the chemical have proven to be very costly.

Another new technique to obtain natural repellents involves identifying the gene or genes responsible for the active ingredients. Those genes are then extracted from one plant and transferred to another. However, many problems arise from this method. Scientists find it difficult to pinpoint the exact gene that fulfills the particular function. In addition, the transfer of genes from one plant to another is exacting and time-consuming work. Another serious consideration is the possibility that the new plant will taste unpleasant to humans as well as to insects.

Disease Resistance

In spite of these problems, other forms of genetic transfer are now under investigation. Current research includes studying plants that manufacture antibodies that fight viruses, bacteria, and fungi. This idea is not new. While working in Mexico in 1944, Borlaug conducted an unsuccessful search for a species of fungus-resistant wheat.

Agricultural scientists have known that wild ancestors of modern crop plants have a powerful resistance to disease. Plant breeders are experimenting with mating wild varieties of crop plants with their domestic kin. By careful long-term breeding, scientists hope to reintroduce the natural resistance of the wild plants and retain the high yields of the domestic plants.

Other new techniques of genetic engineering may increase disease resistance and other beneficial characteristics in crop plants. Experiments are under way to duplicate the natural ability of some plants—such as soybeans—to trap nitrogen. If this ability can be genetically transferred to other plants—such as rice and wheat—the application of synthetic fertilizers would be greatly decreased.

While basic research on nitrogen fixation has been successful in the sense that many of the details of the process are now understood, agronomists have made little progress toward giving more plants such a capability. However, researchers have recently raised hopes for a breakthrough. They have discovered several species of bacteria, in both saltwater and freshwater environments, that can absorb nitrogen from their surroundings and transform it to meet the needs of higher plants. It is possible that these bacteria can be grown in large batches as a way of producing a natural nitrogen fertilizer.

Scientists are also attempting to produce better plants by irradiating seeds with X-rays. The irradiation might result in a beneficial plant mutation. Even though the odds are many thousands to one against such an occurrence, the technique is appealing. The cost of irradiating and then cultivating thousands of seeds is not high. Although most of the plants will be no better—or maybe worse—than their parents, one superior plant could provide a substantial reward to the developer.

For example, during the past few years, Chinese researchers have irradiated immature plant tissue to create mutant rice plants. They have selected the plants that were the hardiest and have propagated them for planting in areas of harsh climate conditions.

Both gene transfer and mutation by irradiation frighten some people. Indeed, some think that such modifications are against nature. Activists warn that farmers are sowing as much as 70 percent of farmland in mutated varieties. Such assertions are intended to frighten consumers. Natural mutations, however, occur all the time. All modern creatures, including humans, are the result of millions of years of natural modifications.

The search for new methods to improve food production is driven by two factors. The most important is that humans must cultivate sufficient food for a growing world population. The second is that insects and other pests that consume the food must be controlled in an environmentally safe manner.

New and experimental methods using genetic transfer in plants and animals provide promise for solving the problem. Nevertheless, scientists are considering other alternatives as well. Fortunately, there are several possibilities.

Preventing Insect Reproduction

Insect birth control is one of the most promising techniques. Various forms of radiant energy—such as X-rays—can be used to sterilize large numbers of insects. The insects can mate, but cannot produce offspring. This technique eradicated the screwworm, a notorious pest that attacked cattle and other livestock in the southern United States. In 1977, after studying the mating habits of this insect, scientists developed a program to sterilize adult male screwworm flies and then release them. Since the sterile insects could not reproduce, the screwworm problem was totally eliminated by 1982. Scientists are currently working on a similar program to eliminate the Mediterranean fruit fly. This pest causes severe problems for the citrus industry.

Scientists studying insect birth control find that many insects require neotonin, an insect hormone, during their life cycle. In order for the insect to reach adulthood, the hormone must be present at certain stages from egg to maturity. However, neotonin is fatal if introduced at the wrong time and so it can be used as an effective insecticide. This method would be relatively inexpensive since each application requires a very small amount of the hormone. Indeed, a large thimbleful of neotonin would

Sterile male Mediterranean fruit flies are released for mating with females that will produce no offspring. (Courtesy of the Office of Communication, U.S. Department of Agriculture)

clear an acre of all susceptible insects; a few ounces would clear an hectare. The hormone is totally harmless to mammals, but unfortunately, it kills both helpful and harmful insects.

Scientists recently discovered a substance produced by the balsam fir tree that closely resembles the chemical composition of neotonin. This natural chemical is not fatal to all insects and might prove to be more useful in selectively destroying just the pests.

Some old-fashioned methods of insect control are still in use. With these practices, all chemicals are avoided. A few farmers continue to pick pests, such as potato beetles, off affected plants.

Crop rotation, used primarily to improve the soil, also helps reduce pest damage. In this method, a field is planted with a certain crop for one year and a different crop the next. Since a favorite food is not regularly available, pest populations do not grow to dangerous size.

Other good farming practices include mechanical mulching— chopping up—or plowing under the leaves and stems of a crop plant after the harvest. The organic materials help the soil retain moisture and regain fertility. Plowing under the debris also removes places where pests can deposit their eggs.

Biological Control

Perhaps the best method to control a pest is to introduce its natural enemy. This method—called biological control or bio-control—is gaining popularity with environmentalists all over the world. Insect pests can be killed by viral diseases, bacterial infections, and fungus attacks or eaten by birds and other animals. Indeed, one species of insects can destroy a different species of insects. A successful program of biological control will often continue to combat the targeted pest without further investment by the grower.

Wasp and looper grub. The wasp inserts its eggs under the skin of the grub. When the wasp larvae hatch, they consume the grub from the inside. (Courtesy of the Office of Communication, U.S. Department of Agriculture)

The classic biological control for both insect and plant pests is the introduction of insect predators. However, the use of pathogens—virus, bacteria, and fungus—is increasing. For example, in the late 1980s, an Australian fungus was introduced into the United States as a means to control grasshoppers. The experiment was successful. Worldwide, about 165 species of insect pests have been controlled by the introduction of their natural enemies.

However, great care must be taken when foreign creatures are imported to control a pest. The story of the cane toad is a famous example of biological control gone awry. In the early 1930s, Australian sugarcane growers were having severe problems. The grub of a native beetle was eating the roots of sugarcane plants. This caused the tall cane plants to fall down. As the beetle infestation spread, the Australian farmers became very worried.

While investigating the problem, an Australian agricultural extension agent heard of a possible cure. Cane growers in Cuba and Central America suggested the use of a native toad as a pest control. This large toad lived in the cane fields and ate many insect pests. In 1935, the extension agent arranged to have some of these toads sent to Australia.

The cane toad, a supposed natural pest controller, turned into a pest in its own right. (Courtesy of Stephen Richards, Vertebrates Department, South Australian Museum)

When the toads arrived, they were put in a special pond to see if they would breed in their new surroundings. They bred successfully, and soon some escaped. The Australians hoped that the toads would head for the surgarcane fields and eat beetle grubs. However, the toads soon discovered that the beetle grubs came out of their tunnels only when the soil was dry. The imported toads—native to a wet, tropical climate—did not like dry conditions. Therefore, they quickly migrated to swampy areas where no sugarcane is grown. Since no beetle grubs could be found in the damp soil, the toads ate everything else. Soon there were millions of toads, but the cane growers did not benefit from the population explosion.

Unfortunately, the toads had no natural enemies in Australia. The cats and snakes that might have killed them were repulsed by the unpleasant, poisonous reptiles. Soon, the toads became pests. Their population had to be controlled by artificial methods such as poisoning. In this case the cure was worse than the problem.

Except for an occasional bad experience—like that with the cane toads—biocontrol is usually successful and generates few serious side effects. However, no foreign species should be introduced into a new habitat without testing and evaluation. In order to anticipate all possible consequences, scientists must conduct extensive and intensive research.

The Middle Ground

Integrated pest management programs provide a middle ground between agricultural programs that employ many synthetic chemicals and those that are totally free from such chemicals. No one claims that IPM is a perfect solution to all arguments between environmentalists and food producers. A major drawback of IPM is the amount of time and money necessary to gain information on the specific biological setting of

The honeybee is one of the beneficial insects that can be killed by careless use of insecticides. (Courtesy of the Office of Communication, U.S. Department of Agriculture)

each case. Pest control managers must be able to determine when to use the available natural defenses and when to use synthetic chemical agents.

The Other Side of the Coin

It is easy to forget that most insects are harmless to humans and, in fact, are beneficial. Obvious examples include silkworms and honeybees. Many insects, like honeybees, serve humans by pollinating flowers. Less obvious is the insects' role as the world's great recyclers. Insects consume a great variety of waste materials and release the materials in the form of natural fertilizers.

Even less obvious is insects' ability to provide nutritious food for many species. Fish, birds, and animals prey on insects that are rich in proteins and vitamins. Insects contribute indirectly to human nutrition when humans eat freshwater fish and birds, such as chickens and turkeys.

Insects also preserve biological diversity. In a field planted with a single crop, as many as 200 different species of insects inhabit each square yard (.8 m^2) of soil.

Finally, the grace and beauty of insects provides humans with great pleasure. No one who has watched a firefly or a monarch butterfly would endorse the elimination of these creatures. Fortunately, the new biological controls will not harm the vast majority of the insect population.

12

A Mature Environmental Protection Agency

President Richard Nixon authorized the formation of the Environmental Protection Agency (EPA) in 1970. Prior to that time, scientists who studied the environment were scattered throughout many organizations and institutions. For example, most regulations having to do with pollution and the enforcement of antipollution laws were within the authority of the individual states. Modest programs of environmental research were being carried out in a few hundred colleges and universities around the country. The most vigorous research programs, however, were being conducted in a variety of government laboratories. Health problems caused by contaminated water or air pollution were the responsibility of the Public Health Service, then a part of the Department of Health, Education and Welfare. The effects of pesticides were being studied by employees of the Department of Agriculture. The policies proposed by some branches of government were, at times, directly opposed to the policies of other branches. Indeed, it was this confusion that helped lead to the creation of a new agency, the EPA, where the whole range of environmental studies could be considered from a single point of view.

Staff personnel and facilities were transferred from several existing agencies and reassembled under the administrative head of the new organization. This administrator was a lawyer named William Ruckelshaus who had served as a government official at both the state and federal levels. He saw the primary role of the EPA as an agency to draft new pollution laws and enforce these laws. At first, Ruckelshaus proceeded in the face of serious disagreements among legislators and special interest groups that sought to influence legislation.

Many people believe that a good way to overcome disagreements and to reconcile opposing views is to assemble all relevant facts concerning the matter at hand. One way to obtain such facts is through scientific research. Thus, the new EPA desperately needed to have its own scientific research capabilities.

President Nixon appointed William Ruckelshaus to be the first administrator of the Environmental Protection Agency in December 1970. He served until April 1973. Then, in 1983, President Ronald Reagan called him back to the agency. His second term as administrator lasted until 1985. (Courtesy of the Environmental Protection Agency)

Building an Environmental Science

A number of scientists and technicians were transferred from other government agencies to staff the new research laboratories. Their first projects were related to assessing the severity of the environmental situation. EPA scientists with a background in agriculture were directed to determine the medical consequences of the use of agricultural chemicals and the routes by which such chemicals enter the general environment. That project and several new ones are still in place. Presently, scientists at the EPA are analyzing the approximately 20,000 pesticides now on the market. These products can contain any combination of the 620 chemicals known to be the active ingredients in such products. This task is very difficult because some 2,000 new pesticides are marketed every year while only a few hundred are discontinued by their manufacturers.

During the first few decades of the Environmental Protection Agency's existence, the larger scientific community believed that EPA scientists were engaged in mediocre research programs. To improve this situation, two special groups were established to ensure more rigorous research and heighten the perception of quality. The Science Advisory Board was established in 1978. The members of the board review the final outcome of each research project and—in conjunction with the research administrators—determine how to use the research in current EPA policy decisions. The Board of Scientific Counselors was established in 1996. This group reviews the EPA's priorities on each research program and advises the scientific staff on the details of research plans. Distinguished members from prominent environmental organizations, private industry, and academic institutions comprise the membership of these groups. The strategy of engaging such groups appears to have succeeded because criticism of the EPA's research has decreased in recent years. There is still room for improvement. Criticism continues from top scientists brought together by the National Academies of Sciences. For example, in 2001, the chair of a National Academies panel testified at a congressional hearing that the EPA was still deficient in some areas and that Congress should encourage the creation of a post for a director of scientific activities at the level of an assistant administrator. Such a person would have direct supervisory responsibility and accountability for the quality of the research.

The New Agenda

The EPA now pursues a wide-ranging research agenda. Some programs—such as habitat restoration—are linked to problems in ecology. Others are connected to both ecology and agriculture. For example, research has proven that wetland habitats serve as buffers that prevent agricultural waste materials from contaminating rivers and streams. Still other programs are narrowly focused on agricultural issues such as the effects of agricultural chemicals on human health.

The work of the EPA is divided among five offices and 10 geographic regions. Four of the five offices are primarily involved with regulatory and enforcement functions, and these offices receive technical support from their own in-house laboratories. Each of the regions has enforcement responsibilities for local monitoring and pollution abatement functions. Most long-range research projects are assigned to the fifth office: the Office of Research and Development.

Units called centers and laboratories are part of the Office of Research and Development. For example, the National Center for Environmental Assessment develops standards of pollution and measures the degree of danger from various pollutants. The work takes place in three sites: Washington, D.C., Cincinnati, Ohio, and Research Triangle Park, North Carolina.

The National Center for Environmental Research and Quality Assurance handles the EPA's research grants administration. Proposals for research from scientists outside the EPA are evaluated on their scientific merit. Most grants are awarded to researchers in colleges and universities.

The National Exposure Research Laboratory includes several divisions that examine agricultural issues. For example, the Ecosystems Research Division in Athens, Georgia, studies agricultural chemicals and the effects of increased atmospheric carbon dioxide on the growth rate of plants.

An environmental technician sampling for water contamination in a stream in Iowa (Courtesy of the Environmental Protection Agency)

The National Health and Environmental Effects Research Laboratory directs at least three subordinate divisions that address the concerns of agriculturalists. These are the Gulf Ecology Division at Gulf Breeze, Florida; the Mid-Continent Ecology Division at Duluth, Minnesota; and the Western Ecology Division in Corvallis, Oregon. The three divisions have the responsibility for studying the contamination of bodies of water by agricultural runoff—particularly the waste products from concentrated animal feedlots.

At present, the EPA is leading the world in the study of the strange microorganism, *Pfiesteria.* These amoeba-sized creatures live in rivers and coastal waters and exhibit at least 26 different life stages or physical forms. In certain stages, they release a poison that can cause sores to erupt on the skin of small fishes. In others, they are actually environmental benefactors because they consume harmful bacteria. In still different forms, the microorganism appears to lie dormant on the bottom of the water and to await some kind of stimulus to become active. Some scientists suspect that the main stimulus could be an abundance of nitrogen or phosphorous—accumulated from agricultural fertilizers or animal waste and washed into streams by rainwater. However, this connection has not yet been proven.

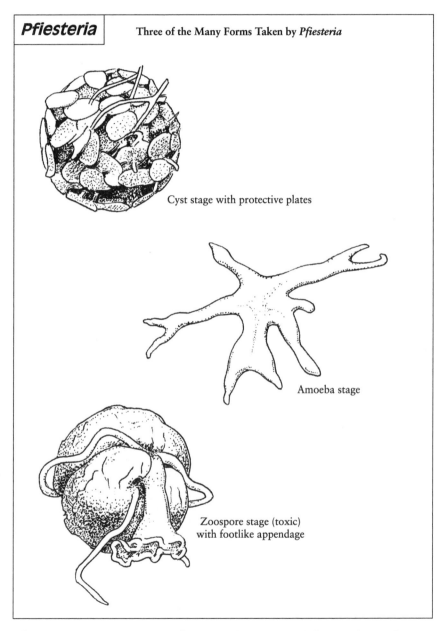

Pfiesteria

Three of the Many Forms Taken by *Pfiesteria*

Cyst stage with protective plates

Amoeba stage

Zoospore stage (toxic)
with footlike appendage

Pfiesteria *is a microorganism that can take on several distinct forms during its life; one such form is toxic and can cause skin lesions upon exposure.*

One of the principal sites for the EPA's *Pfiesteria* research is the Center for Applied Aquatic Ecology at North Carolina State University in Raleigh, North Carolina. However, many other organizations are involved. At the federal level, the National Institute for Environmental Health and the National Oceanographic and Atmospheric Administration are doing their own studies. At the state level, the natural resources departments of all states in the South Atlantic region are involved in *Pfiesteria* research. Various departments of the University of Maryland in College Park are also closely involved in this investigation.

Finally, the EPA has recently shown a special interest in the exhaust products of diesel engines. Diesel exhaust tends to be highly "sooty." That is, it contains many small—even microscopic—particles. Some of these particles are made of

A small freshwater fish with a lesion on its underside thought to have been caused by microscopic Pfiesteria (Courtesy of the Environmental Protection Agency)

unburned carbon and others contain tiny mineral grains like powdered sand.

The diesel engines in cars, trucks, and buses are already under some regulation but the so-called off-road or stationary diesels have generally escaped restriction. Many stationary diesels are used on farms to drive water pumps and emergency electrical generators. In addition, some large tractors are diesel-powered. The EPA is planning a research program related to the suppression of particles emitted in diesel exhaust.

While it seems unlikely that farmers will welcome any new restrictions on their use of diesel engines, the EPA may have a recommendation that would make such changes more acceptable to the farm community. The agency is considering the possibility that some vegetable oils could be added to the traditional petroleum oils to make a new and better diesel fuel. In that case, the new diesel fuel would include a renewable energy source, and the use of vegetable oils would provide a new market for an important agricultural product.

13

Ecology and Agriculture

Ecology is a branch of the life sciences that pulls together elements from a variety of other fields. Ecological research differs from that of other basic sciences because it is rarely experimental and has few practical applications. Chemistry, in contrast, is predominantly an experimental science that derives many products (such as medicines) from research. In some ways, ecology resembles the older science of natural history that was dedicated to the discovery and classification of individual plants and animals. However, ecologists study living things in communities rather than single examples of a species. Ecology is also like astronomy because its study can lead to the understanding and prediction of natural events but cannot provide the means to control the events. In short, ecology is based on the careful observation of plants and animals, but ecologists have no inclination to intervene in the natural order of events.

Advances in the science of agriculture have generated improved practices—or at least, practices that are technically efficient. Advances in the science of ecology have yielded principles and rules of behavior and have provided agriculturists with a particular set of ethical values. However, since agriculture is primarily based on practical applications rather than philosophical concepts, the connection between ecology and agriculture has been limited.

Soil Erosion

The need for a comprehensive value system to guide public policy in agricultural issues appeared in the early days of the 20th century. At that time, Hugh H. Bennett, a young soil scientist working for the U.S. Department of Agriculture, became alarmed by what he perceived as the danger of soil erosion. Bennett noted that a large decline in crop productivity had been seen in areas of the country where erosion was out of control. He began to campaign for research that could prevent such soil losses.

In spite of Bennett's effectiveness as a propagandist, the problem of soil loss continued to worsen. In 1914, the stresses of World War I generated top prices for wheat. Land that was best suited for grazing cattle was plowed and planted in wheat. When prices dropped at the close of the war, the land was turned back to grazing, but it was too late. Grasses did not grow fast enough to provide sufficient cover for the land, and the herds of cattle disturbed the thin layer of soil. When

A farmer and his two sons rush for shelter from an Oklahoma dust storm in 1930. (Courtesy of the Photographic Service and the Farm Security Administration of the U.S. Department of Agriculture)

drought conditions arrived in the 1930s, ordinary wind storms picked up the soil and soon became dust storms of giant size. Farming was next to impossible because the once fertile soil had been blown away. For several years, farms in the center of the United States became part of the so-called Dust Bowl.

By 1935, these conditions were seen as a national crisis. The federal response was to create the Soil Conservation Service (SCS). Bennett was assigned as chief of the service and served until he retired in 1951.

Others shared Bennett's concerns about the environment. Paul B. Sears, who received a graduate degree in ecology from the University of Chicago, finished his book, *Deserts on the March,* in the same year that the SCS was founded. His book— concerning the loss of land fit for cultivation and the spread of desert areas—provided a link between ecology and agriculture. Top government officials under President Franklin D. Roosevelt, such as Rexford Tugwell, a prominent Roosevelt advisor, were encouraged to join the conservationist movement. Tugwell served as Under Secretary of Agriculture from 1934 to 1937.

Many other opinion leaders joined the cause of conservation and sustainable agriculture. Reformers advocated the use of leaves and stems left behind after the harvest to make mulch for fertilizing crops. The scientific basis for this and other recommended practices was thin. However, the value system behind such ideas had broad appeal to all citizens. Urban dwellers were convinced that they could improve their lives by fertilizing their organic gardens with mulch from lawn clippings.

Louis Bromfield

A writer named Louis Bromfield (1896–1956), born near Mansfield, Ohio, was among the most articulate and influential converts to conservationism. Bromfield began college at Cornell University in Ithaca, New York, and his study of agriculture there allowed him to appreciate the principles of farm management. However, at the urging of his mother, he soon transferred to Columbia University in New York City and began to study literature. In 1917, after one year at Columbia, he dropped his studies and enlisted in the French Ambulance Service in support of the Allied war effort. Bromfield served with distinction and developed a strong fondness for French culture.

After the war ended in 1918, Bromfield returned to the United States and began his career as a writer. He was successful as a novelist and a Hollywood screen writer and in 1927 won a Pulitzer Prize for one of his novels. Soon, Louis Bromfield was financially able to return to France with his wife and children, and he purchased a house just outside of Paris. At the outbreak of World War II in 1939, concerns for the family's safety forced their return to the United States and to his hometown of Mansfield, Ohio.

While renewing his acquaintance with Mansfield and the surrounding area, he was disappointed to see that once prosperous farms were now rundown and the land eroded. Gullies pierced some fields, and a pond in which he had swum as a boy was clogged with green scum. After looking about, he decided to restore a large, once productive farm by using established ecological principles. Bromfield bought 1,000 acres of rolling hills and called it Malabar Farm.

He restored the land by filling in gullies and replanting the hills with grasses that would hold the soil. After a few months of intense planning, he began building a large house to accommodate his family and his associates. Bromfield was soon able to demonstrate the worth of organic practices, and the farm attracted many visitors. He was proud to show that the once polluted stream now ran clear and could support bass and other fish. In addition to dairying and raising field crops such as corn, he planted truck crops such as carrots and lettuce and offered surplus produce for sale at a roadside stand. All

Louis Bromfield studying the nature of the soil in an erosion cut on Malabar Farm before his restoration was started (Courtesy of the Ohio Historical Society)

the Malabar Farm vegetables were truly organic. Bromfield died in 1956 at the age of 59. Malabar Farm was deeded to a trust that was committed to operating the farm using his ecological principles. Unfortunately, those left in charge had neither the charisma nor the energy of Louis Bromfield. In 1975, the mortgage holders were preparing to foreclose. However, since the mortgage was held by another philanthropic trust, an amicable outcome was arranged, and the property was deeded to the state of Ohio. In return, state officials promised to maintain the ecological demonstration and ensure permanent public access to the estate.

The Bromfield experiment was only one small part of the soil conservation movement. Other activists such as the Rodale family—famous for their modern farming techniques—and another successful writer, Edward Faulkner—son of the world-famous author, William Faulkner—continued to push for the acceptance of ecological values and the success of the conservationist movement.

Expanding Acceptance

Today, many farmers follow ecological principles without knowing the historical background of the movement. A farmer who allows a field to be planted in corn one year and in clover the next is helping to restore the field's natural fertility. He is using the principle of crop rotation to maintain soil fertility—one type of recycling. If the farmer allows his farm animals to graze in an uncultivated field, the process of soil restoration has been further strengthened.

Other ecological principles are followed when a farmer plows a hillside grain field so that the furrows cut across the face of the hill rather than up and down the incline. Rainwater cannot wash the topsoil down the hill because the furrows act like miniature dams and retain the soil and water.

Fifty years ago, the Department of Agriculture had no confidence in the new science of ecology. Many top officials thought conservationism was irrelevant. Today, the public statements of government officials reflect a sharp change in attitude. Activists hope that their new acceptance of ecology will continue to grow.

14

Environmental Update

Since the EPA was established in 1972, the scope of environmental protection has expanded, and many new laws have been passed. The EPA has succeeded in reducing air and water pollution with its new regulations. The federal government, working with local communities, has improved wastewater treatment plants, and waterways are now safer and cleaner. Nuclear power, wetland preservation, solid waste recycling, endangered species, and toxic chemicals in the air and water are now monitored.

Today, the determination to protect the environment is global in scope. Many local regulations have been taken over by national agencies. National regulations have been expanded into international agreements. Still, the struggle to achieve a safe environment is far from over.

What's Next

The dispute between hunger fighters and environmentalists will continue—but in a less strident manner. Many of the disputed issues have already been partially resolved. The scientific community now agrees that excessive use of synthetic pesticides is a losing strategy. Integrated pest management (IPM) has been widely promoted as a good compromise. Food production and environmental protection are both served by the use of IPM. Even in some nonindustrialized countries, this new approach is now the official doctrine.

The movement toward IPM does not mean that the problem of environmental abuse has been completely solved. Many farmers around the world do not have the knowledge or skills to follow IPM practices. Even in the United States, overuse of pesticides can occur on corporate farms that employ the most advanced technology. Such instances reflect how short-term profit can outweigh long-term environmental good.

Further Productivity Gains

Without a doubt, the agricultural green revolution will continue. Breeders will develop new strains of corn, wheat, and rice that can resist pests and diseases. Unfortunately, after five or six years, the new strains may fall prey to adapted pests or new forms of disease. The breeders will then begin again.

Each new strain will possess genes that provide high-yielding plants. Today, these genes are found in almost all cultivated crops of wheat, corn, and rice. Some believe that scientists have exhausted all genetic possibilities to increase crop yields. However, future research may produce a plant capable of still higher yields. Meanwhile, breeders are striving to improve other factors, such as pest resistance.

Scientists are also experimenting with grains such as oats and sorghum. Various breeding techniques are being tried, and some success has been achieved in crossing wheat and rye. Breeders hope to develop a new plant that will exhibit the hardiness and short growing season of rye and the high-yield characteristics of wheat.

Other staple crops such as soybeans, and root plants such as potatoes, yams, and cassava have been the subject of recent experimentation. In addition, the full range of table fruits and vegetables are being crossbred or hybridized to develop superior products.

New Research Fronts

Research points out the possibility of producing food from perennial plants. Today, bulk foods such as grain come from annual plants that need to be reseeded each year. However, many of these annuals have cousins that survive from year to year without replanting. For example, a perennial cousin of corn is called gama grass. Usually, this plant does not produce much grain. All its energy is used to build a root structure that can survive the winter months. However, a common mutant variety of gama grass puts much more of its energy into seed production. This gama grass gives four times as much grain as the nonmutant variety.

Agricultural scientists have developed a method of cultivating perennial grain plants called "natural systems agriculture." This method requires several different species of perennials to be intermixed in a field. In addition to the grain species, scientists add a perennial bean species to capture nitrogen for natural fertilization. A natural insecticide producer such as chrysanthemums can also be a part of the mix. Several benefits flow from this technique. Soil erosion is prevented because fields containing perennial plants are not plowed. The cost of agricultural chemicals is avoided because the plants have a natural supply of fertilizer and insect repellent. Also the consumer enjoys the advantage of chemical-free food.

Scientists estimate that long years of research are needed to refine the technique. Among other concerns, they must solve the problems involved with mechanized harvesting of a diverse crop of plants. Most agree, however, that the success of natural systems agriculture would prove to be a great benefit to humans.

Food Production from Vats

Another technique to increase plant productivity involves the study of relatively simple cell chemistry. All the chemicals needed for balanced nutrition can be manufactured by each unspecialized plant cell. The unspecialized cells of grain seeds, for example, produce starch, sugar, cellulose, protein, and vitamins. Unspecialized plant cells can manufacture this

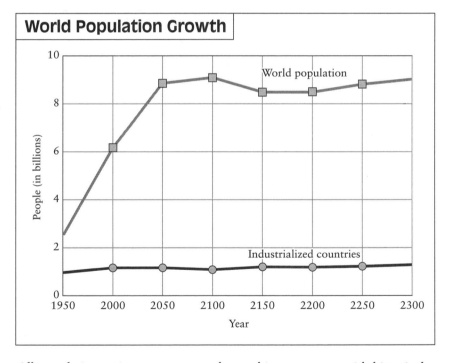

World Population Growth

All population estimates are somewhat ambiguous—even with historical data. The best current guesses project world population growth to more than 10 billion—but the growth should level off around the year 2200.

nutritious mixture even if they are isolated from the parent plant.

In a complete, mature plant the cells become specialized, and most of them become part of a leaf, stem, or root. Consequently, much of the plant is not usable as food. Scientists have found that each of the unspecialized cells can be grown as a mass in which all cells are exactly alike. This suggests that unspecialized cells might be a more efficient source of food. Scientists are studying this possibility and are attempting to grow wheat, corn, and rice cells in large vats.

In fact, yeast cells have been grown in such conditions for hundreds of years. The cells manufacture various proteins and produce alcohol as a by-product. The yeast cells are processed into food, and the alcohol is used as a much needed gasoline extender. When yeast or other microorganisms are grown in this way, the process that takes place is fermentation. This same process, using yeast and wheat flour, is what makes bread dough rise. Other foods that are produced by the actions of microbes include yogurt and all the many varieties of cheese.

The production of food in vats has serious limitations, however. A basic problem is providing the necessary nutrients for cell growth. Also, the process of transforming the mass of cells into an attractive, appetizing food could be costly. Lastly, some people would consider the artificially produced food to be unwholesome—no matter how tasty and nutritious. Several lines of research have opened the prospect of overcoming these limitations. One line is biological and involves the genetic modification of the microorganisms that carry out the fermentation process. The genetic modification changes the proportions of the products of the process. For example, a goal might be to produce a particular protein or one of the amino acids that form proteins. The genetic makeup of the microbe contains the plan for its protein production, so if it can be modified, the proportions can be changed. In the fermentation batch, many different proteins will be generated, but with a genetically modified microbe, the proportions would shift in favor of the most valuable product.

A second line of research is directed toward the automation of the process. The main idea is to automatically detect the needs of the microbes for nutritional materials and meter the provision of these materials to exactly match those needs.

15

Genetic Modification

In 1984, 30 years after Gordie Hanna developed a tomato that could survive mechanical harvesting, another tomato project was launched in the central valley of California. This new endeavor was preceded by the hard work of a small group of plant scientists, biochemists, and molecular biologists. They

Tomato being tested for firmness (Courtesy of the Agricultural Research Service, U.S. Department of Agriculture)

had come together in the early 1980s to begin an enterprise that would bring genetic engineering to bear on some large-scale agricultural problems. The scientists hoped to develop a new strain of cotton that would be resistant to particular herbicides. If such a strain could be realized, cotton plants would be unaffected by the weed-killing chemicals.

In the middle of the project, the scientists found an opportunity to achieve an even greater economic success. The venture involved the development of a new type of tomato. In contrast to the rough-tough varieties developed by Hanna for canning, the new tomato was destined for the fresh tomato market—eventually to be quartered or sliced for a salad. Hanna's tomatoes could be picked ripe and red because they would be cooked and peeled within a few hours. Their bright red color was necessary to give the right hue to ketchup and spaghetti sauce. However, salad tomatoes had a long way to travel before they would be sold to a consumer and eaten raw.

Tomato Culture

During the long cold-weather season, most salad tomatoes are grown in Florida, the other Gulf States, Mexico, and California. They are picked green so they will survive the long trip north to the large, urban market centers. When the tomatoes arrive at the markets, they are still green and hard.

Batches of tomatoes are then placed in special storage places and exposed to ethylene gas. Ethylene gas is a natural product which supports the ripening process of most fruits. Warehouse workers use an abundance of artificially derived ethylene to control the ripening of the green tomatoes. Unfortunately, as these tomatoes turn red and begin to soften, their texture and taste are inferior to the fresh, home-grown garden tomatoes.

The goal of the new project was to develop a tomato that could be grown in large quantities and transported hundreds of miles but that would retain the taste, texture, and appearance of the home-grown variety. Hopefully, the popularity of fresh tomatoes would allow the developers of such superior fruit to make millions of dollars.

The plant scientists reasoned that a study of pectin—the natural plant product that helps tomatoes to remain firm—might reveal the answer to their tomato problems. In nature, an enzyme dissolves the pectin as the tomato softens and becomes fully ripe. If the pectin could be retained in the plant throughout the ripening process, the tomato would not soften. The fruit could be left on the vine until it was a deep red but remain firm enough to travel long distances without becoming mushy. Furthermore, if the fruit stayed on the vine until shipped, the plants might be able to form the sugars and flavoring factors that make tomatoes taste good.

The scientists knew that a particular enzyme acts to dissolve the pectin. This enzyme is created inside the plant cells when the tomato has attained its full size but not its red color. The enzyme is known to be produced by the action of a particular gene, and the DNA sequence of that gene had been recently discovered. The researchers reasoned that if they could stop or reduce the action of the gene, the pectin would remain intact in the fruit.

Changing Genes

They began to study the possibility of developing an artificial gene that would slow the production of the enzyme that dissolves tomato pectin but would be accepted by the tomato's enzyme manufacturing system. Of course, to accomplish this trickery, the fake gene would need to be very similar to the real gene. The scientists were clever. They reversed the natural order of the DNA sequence to build an artificial—but hopefully acceptable—gene.

Now the scientists addressed the problem of getting the fake genes into the cells of the tomato plants. Research had shown that an otherwise harmless microbe made a practice of attacking plant cells by injecting its own DNA into them. The microbial DNA led the plant cells to produce proteins that were beneficial to the microbe but harmful to the plant. Using some elaborate and delicate manipulations, the scientists were able to remove the harmful genes and implant their fake genes into a large batch of the bacteria. The bacteria were then added to a liquid medium that contained very young tomato plants—clusters of fragile cells that had just emerged from tomato seeds. Many of these thin-skinned cells were attacked by the bacteria and injected with the fake genes. The cells were carefully nurtured and grew into mature tomato plants. If all went well, the succeeding generations of seeds would contain the artificial gene.

While working on the technical difficulties, the researchers became aware of many business problems. The first concerned two competitors. Scientists at the University of Nottingham in England were working on fake gene substitutions and specifically on the anti-pectin gene in tomatoes. Their lawyers in America were seeking a U.S. patent for the artificial genes that they were developing. The California group and the English group were in a race to see who would be awarded the exclusive use of such genes. Both groups were well-subsidized. The English group was financed by the Unilever Corp., a gigantic international food and cosmetics company. The Americans were supported in part by the Campbell Soup Company.

As a secondary distraction, scientists at the University of California at Davis were also testing the effects of reverse order genes. They were not interested in patents or commercial advantage, but they were eager to publish the results of their research. If the Davis scientists were the first to publish, this information could block the award of patents to other contenders, and the reverse order technique would remain in the public domain.

Another set of problems concerned the safety to consumers and the nutritional value of a genetically modified food. At the time, there were no specific standards for the safety approval process on genetically modified foods. No determination had been made as to whether the U.S. Food and Drug Administration or the Department of Agriculture was to be in charge of protecting the public.

After this issue was resolved and the U.S. Department of Agriculture was given the lead role, many uncertainties remained. This situation persuaded the leaders of the California group to make public all the information acquired in the developing and testing of their genetically enhanced tomato plants. This tactic meant that they accepted the risk of presenting both positive and negative findings to their future customers. For example, one study to ensure the safety of the new tomato involved feeding the tomato to laboratory rats. Some rats were fed the new tomato, some were fed ordinary tomatoes, and a third group was fed standard laboratory rations. After several weeks on the feeding program, the researchers discovered that some of the new tomato–fed rats had developed sores in their stomach linings. This was a disturbing find. Nevertheless, the researchers made these results public. Almost immediately, they conducted the same experiment on a much larger scale. This time, the results showed that the rats fed on all the feeding programs developed about the same number of stomach sores. The only logical conclusion was that the rats were equally prone to stomach sores and that the new tomato was no more to blame than the other foods.

Official Approval

The openness of their communication led to both positive and negative responses. The California team was criticized by people who were opposed to genetic manipulation of foods.

However, government regulators were convinced that the research project was honest. In fact, the strategy led the regulators to declare that the new tomato was safe and nutritious. Moreover, they decided that the new tomatoes were not basically different from those developed by traditional cross-breeding. This decision seemed to indicate that even more exotic genetic manipulations, such as transferring genes from one species into another, might be acceptable in the future.

Objective observers from outside the government and the food industry, such as members of the National Academy of Sciences, sought to find a fair balance between the interests of groups for and against genetic manipulations. That effort continues.

To the Marketplace

After approval of the new tomato, other difficulties arose. The developers now needed to decide how to produce large quantities of the tomatoes and transport them from farm to market. The development costs had been in the millions of dollars, and the researchers needed to repay the investors. They also required money to support the sales force necessary to persuade supermarket managers to stock the new tomato, which would sell for a higher price than the public was accustomed to paying. This price reflected the high expenses and allowed for the company to make a profit. At first, the developers were convinced that the availability of a red, ripe, juicy tomato during the fall, winter, and early spring would bring a premium price. However, there was one final problem. The developers had neglected to test the old adage—"The proof of the pudding is in the eating." They had paid little attention as to whether a superior taste and texture would result from keeping a tomato on the vine until it was fully mature. In the early stages of development, there had been some hesitancy to sample a product that had not been certified as safe. In the few tests that were conducted, the tasters were told to chew but not swallow the fruit. In any case, the developers had completely neglected the crucial judges—chefs, supermarket managers, and ordinary people who bought food for their families. In fact, when the time came and tons of tomatoes were delivered to the stores, the new tomatoes were judged as acceptable but not tasty enough to command a premium price. This judgment, combined with the fact that the developers had no experience in growing and selling large numbers of tomatoes, forced the group to sell the business. It was purchased by the Monsanto Company but even this giant agriculture firm soon gave up the dream of offering a superior winter tomato. No such product is now listed on their Web site.

New Horizons

While the developer's dreams were never fulfilled, several lessons had been learned. First, the commitment by the scientists to communicate all of their research findings encouraged the acceptance of their work by the scientific community, government officials, and ordinary citizens. This tactic had proved successful.

Second, it became apparent that improvement was needed in the procedures for government review of new foods. Today, the basic practice for all government actions is to post information for each case in the *Federal Register*. This information includes a summary of evidence, the substance of a forthcoming decision, or information on a change in regulations. The *Register* is similar to a newspaper except that the articles tend to be of interest only to the people or industries that are affected by the government's actions. Members of the general public do not often read the *Register* even though it is freely available in many libraries. However, organizations acting for special interests or certain segments of the public maintain staff members who are assigned to scan the *Register* on a regular basis. Then, when appropriate, the organization can produce a response—such as a strong objection to the proposed changes. This means that those citizens who are members of activist organizations are likely to have more voice in the governmental supervision of science and technology than nonmembers of such organizations.

Lastly, the whole episode tended to show that few, if any, commercial companies can survive solely on the strength of their laboratory work. High science or high technology might produce miraculous products, but without the capability to obtain the necessary financing and the expertise in manufacturing, handling, storage, accounting, sales, and customer relations, success will be elusive. Perhaps that is one reason why pure science tends to thrive mainly in academic settings.

In any case, the lessons were not lost on the officials of the Agricultural Research Service, the scientific arm of the Department of Agriculture. Their scientists formed a team with scientists from the University of California at Berkeley to try an alternative tactic in the production of a superior salad tomato. Instead of trying to slow the destruction of pectin in a tomato, the new approach would halt the production of ethylene and therefore slow the ripening of green tomatoes. The production of ethylene—as in that of pectin—is the responsibility of a single gene. If the production process could be fooled by a fake version of that gene, little ethylene would be generated. The fruit could remain on the vine and stay green during the maturation process necessary to produce the sugars and flavorings that make a good tomato. The produce could be shipped green just as winter tomatoes are now. After reaching the chosen destination, the tomatoes could be exposed to large volumes of ethylene and quickly attain the desirable red color. The dream to achieve a tasty tomato lives on.

Moreover, the government has moved to ensure that business expertise will be included in the production of the newest super tomato. They have licensed the product to a major vegetable marketing firm that will oversee the planting and handling of the produce. If the venture is successful, the government and the marketing firm will share the rewards. If it fails, the government loses only the cost of the research.

16
Government Programs in Agricultural Research

The main theme of the studies performed by the research wing of the U.S. Department of Agriculture (USDA) has been, for many years, the technical efficiency of agriculture. This theme has been realized by the development of new varieties of crops, cattle, swine, sheep, and poultry. In addition, research programs have determined the best seeds for particular soil and climate conditions. In the past, some research also has been directed to the development of farm machinery and a host of laborsaving devices such as milking machines. Critics of the USDA's research efforts have focused their attention on projects such as massive irrigation and sprinkler systems that appear to favor corporate farm managers

Soy plants ripe in the field and ready to harvest. Most soy now grown in the United States is genetically altered. (Courtesy of the Photography Center, Agricultural Research Service, U.S. Department of Agriculture)

The Wallace Center in Beltsville, Maryland. The headquarters and main experimentation site for the Agricultural Research Service. (Courtesy of the Photography Center of the Agricultural Research Service, U.S. Department of Agriculture)

and owners rather than the independent farmer. However, in recent years the federal government has decreased the funding for such projects.

The success of the new varieties of corn, wheat, and rice has inspired further studies on such staples as soy beans, sorghum, millet, sugarcane, cassava, and plantains. Several of these crops are native to the tropics, and the expansion of the USDA's scientific research agency, the Agricultural Research Service (ARS), to international locations is allowing farmers in other countries to benefit from the research. The development of new peanut varieties also fits this pattern. Research on traditional orchard and garden crops—such as apples, pears, and peaches—has been expanded in recent times to include such tropical species as cacao, mango, papaya, and lychee.

Another effort is the development of livestock and crop varieties that have a high natural resistance to various sources of stress. For example, the ARS helped American farmers import Brahma cattle from India. Crossbreeds of Brahmas with Hereford cows produce cattle that can resist heat stress and are relatively inexpensive to fatten for market.

Disease Prevention

Preventing plant disease is also part of the ARS agenda. Wheat rust—a fungus infection that can destroy whole fields in a brief span of time—is the classic villain of plant disease. In northern California, Oregon, and Washington, wheat rust has consistently reduced yields by 20 percent per year. A related fungus has begun to attack barley, and another related disease called smut has invaded wheat fields in the northwestern part of the country. A disturbing complication is that these fungi adapt very quickly to countermeasures, and wheat rust has developed a gradual immunity to fungicides. Therefore, either stronger chemicals or new formulations are needed from year to year. Wheat rust also seems to overcome a plant's built-in resistance to disease. Researchers at

Magnified image of a stem of wheat infected with the rust fungus (Courtesy of the Indiana Agriculture Experiment Station and Purdue University)

A wheatfield flattened by an infestation of rust microorganisms. Farmers in the northwestern states have experienced losses of as much as half their crop of wheat because of this scourge. (Courtesy of the Agricultural Research Service, U.S. Department of Agriculture)

the University of Washington at Pullman have found varieties of wheat that are rust-resistant, and when crossed with high-yield varieties, the resistance can be passed to successive generations. However, the immunity lasts for only a few seasons. Soon, the rust fungi returns and is able to resume its devastation of the wheat fields. The next stage in the wars on wheat rust and smut will likely include attempts to develop longer-lasting immunities by means of more aggressive genetic engineering.

Pest Control

Such tactics have been used in the struggle to control the insect pests of food crops. In order to achieve reliable pest protection, researchers have seen a need to go beyond classical breeding techniques and into genetic engineering. This decision has, in turn, raised new questions about environmental integrity. The case-in-point used by environmental activists is the implantation of a gene from a bacterium into corn. This bacterium provides the corn plant with the blueprint for a protein that acts as a larvicide (larva-killing material) that is poisonous to moth and butterfly larvae. The goal was to make corn resistant to the larvae of the corn borer caterpillar. Environmentalists saw the development as a potential

Monarch butterfly larvae feed exclusively on milkweed plants. The adults migrate thousands of miles to overwinter in a subtropical climate where their nesting areas are endangered. (Courtesy of the Agricultural Research Service, U.S. Department of Agriculture)

threat to a host of innocent insects, singling out the monarch butterfly as a possible victim. Their scenario emphasizes the fact that the larvicide is incorporated in all the plant's cells, including its pollen. While butterfly caterpillars feed exclusively on milkweed plants and never attack corn, the pollen from corn is highly mobile and is typically blown over large areas. The milkweed plants upon which the butterfly larvae feed could be within those areas. If butterfly larvae accidentally ingest some of the pollen, they might be poisoned. Most entomologists find this scenario to be improbable. Indeed, butterfly enthusiasts

Corn earworms eat any green or ripe produce but like to hide in the corn silk at the tip end of a ripening corncob. (Courtesy of the Agricultural Research Service, U.S. Department of Agriculture)

interested in preserving the species are far more concerned about the loss of the monarch's winter habitat in Mexico than about transgenic larvicides.

Science-oriented environmentalists in Europe, however, raise a more pointed objection. They see the possibility that the larvicide could damage other beneficial insects. Specifically, they note that lacewing bugs feed on corn borer larvae and therefore act as a natural control on that predator. The lacewing bug may be harmed if it eats corn borer larvae that have ingested transgenic corn containing the bacterial toxin. Studies show that the mortality rate of lacewings and other beneficial insects, such as ladybirds, is higher after eating pests that contain a dose of larvicide.

While the use of natural enemies to control pests has expanded, chemical methods of control are still in use. However, much more attention is being paid to limiting the amount of the chemical application. Special equipment has been developed by Agricultural Research Service scientists to limit the flow of herbicides drop by drop. The chemical is applied only to the narrow zone between rows of crop plants.

Similarly, aircraft spraying large areas do not always apply insecticides. New fogger machines can apply the chemical in a much more restricted manner. However, genetics may be the key to the gradual elimination of most of the toxic chemicals used in agriculture.

The larger, brownish lacewing larva is swallowing a whitefly nymph that is barely visible as it goes down. (Courtesy of the Agricultural Research Service, U.S. Department of Agriculture)

Other Genetic Modifications

While the threat to butterflies is not very serious, and the threat to beneficial insects is debatable, it is clear that the whole idea of transgenic technology is unnerving to some people. These critics find it unnatural to take a gene from one species and introduce it into the genetic complex of a different species. The European Union (EU) now requires all transgenic seed and foods, such as fruit, to be clearly labeled. Indeed, there is a strong prospect that the EU will forbid the import of such foodstuffs. Nevertheless, it seems likely that genetic engineering will become a prominent factor in the improvement of plant resistance to disease, drought, predators, and other

Sampling the insect population in an Iowa wheat field with a vacuum sampler (Courtesy of the Agricultural Research Service, U.S. Department of Agriculture)

sources of stress. Research in this area is underway. Specifically, the genetic resources of hardy plants, such as the wild mustard (*Arabidopsis*), are being investigated to determine which of their genes give them a resistance to drought.

Sustainability

Although attempts to move genes from one species to another have generated criticism, American agricultural scientists have

A newly designed insecticide fogger for use with orchard crops. It provides superior coverage of the treetops while using as little as one-third the amount of chemical. (Courtesy of the Agricultural Research Service, U.S. Department of Agriculture)

disregarded most complaints about transgenic plant developments. However, the basic critique—that much of agricultural technology may be harmful to the environment—has begun to instigate reforms. Concurrently, the Agricultural Research Service (ARS) is evaluating alternative methods to reduce soil erosion. One such alternative is to stop plowing and disturbing the soil altogether. The no-till method—in which farm fields are left in a natural state—results in a recycling of plant wastes, reduction in the use of insecticides, and reduction of unnecessary disturbance to the natural habitat.

A related alternative is the ridge-till method. This method requires the use of special equipment to form the ridges, which act as small dams that keep the soil and water from sliding down the hill.

Such alternatives represent major deviations from the traditional theme of agricultural research—the attainment of maximum crop yield. Although they have the advantage of sharply reducing soil erosion, the restrictions imposed by the no-till method lower crop yield by 10–15 percent and that of the ridge-till method by 5–10 percent. Critics of these methodologies are unhappy with the loss of productivity.

Regardless of their critics, the ARS has officially adopted an ecological orientation. This change is apparent at all levels of the research community. The University of California at Davis, once the target of environmental activists, has now adopted a curriculum that is based explicitly on ecological ideas. The university has renamed their agriculture school the College of Agricultural and Environmental Sciences. Within the college, the Graduate Group on Ecology—a new, advanced degree program—has enrolled more than 200 students. A major section of the college curriculum is now labeled as "agricultural ecology."

Another example of this trend is the program being carried out at the Dale Bumpers Small Farms Research Center in Booneville, Arkansas. Named after a recently retired U.S. senator from Arkansas, workers at the center study three important

This ultra-low-volume herbicide applicator developed by plant physiologists is intended to reduce the use of agricultural chemicals. (Courtesy of the Agricultural Research Service, U.S. Department of Agriculture)

areas of family farming. Their studies in agroforestry focus on increasing the amount of reforestation and decreasing the time farmers must wait to recover their financial investment in such endeavors. The development of faster-growing pine trees and the possibility of increasing a market for walnuts and pecans are two of the options under consideration.

Other areas of research include small farm agronomy. Specifically, the researchers are testing the possible advantages of sowing pastures with tall fescue grass or Bermuda grass to provide a fast-growing food source for cattle and a soil-holding root system to forestall erosion.

A third area, integrated systems research, is focused on the prospect of income generation from environmentally approved practices. An example of such a practice is the use of riparian buffers, strips of fallow (unfarmed) land that lay between farm

Corn is planted in the residue of leaves and stems of the prior crop. This is no-till cultivation. (Courtesy of the National Resources Conservation Service, U.S. Department of Agriculture)

fields and streams. The strips protect the streams by safely absorbing the farm waste or fertilizer runoff. The goal is to find the least expensive way to establish such protective zones or, even better, to find ways to generate income from them. Planting garden vegetables in such strips might be one way to extract extra benefits from the ecologically sound practice of protecting a stream.

Local Programs

In 1994, the U.S. Congress passed a statute that provided for the reorganization of the research arms of the U.S. Department of Agriculture. Agricultural extension work, the research conducted by the nation's land-grant colleges

Soybeans are planted on ridges. In the swale between the ridges, waste from the prior crop holds down the growth of weeds. (Courtesy of the National Resources Conservation Service, U.S. Department of Agriculture)

and universities, and the agricultural experiment stations were consolidated into a new unit, the Cooperative State Research, Education and Extension Service (CSREES). One objective of the reorganization was to strengthen the linkages with local citizens through a strong carryover of the historical success of the extension arm. Through the work of extension agents in virtually every county in every state, this department of the federal government has had extensive personal contact with its clients. No other government agency has such a close relationship with its beneficiaries.

A contour-plowed field, so treated to reduce the erosion from rainwater. (Courtesy of the National Resources Conservation Service, U.S. Department of Agriculture)

One important result of this long-standing level of interaction has been the enormous increase in agricultural productivity. The extension program has been so successful that its reach now includes all rural residents—both farming and non-farming families. Extension agents offer advice on diet, sanitation, and general home economics and promote organizations such as 4-H clubs that allow rural children to participate in enjoyable and constructive projects. However, the prime role of the extension agent is to report the latest developments in the field of agriculture. This information includes ways to resolve farming problems, discussions of new farming techniques and machinery, and the results of agricultural research. The extension work is coordinated by officials in the 77 land-grant institutions that are situated in the United States and other U.S.

jurisdictions—such as universities in Guam, American Samoa, the District of Columbia, Micronesia, Puerto Rico, and the U.S. Virgin Islands. Usually, the states administer their county extension operations from offices at the land-grant institutions. Often, the agricultural experiment stations are also closely allied with these same institutions. The research programs of these institutions are tightly coordinated because the federal government helps provide funding for their investigations. The land-grant colleges and universities receive their main support for educational functions from state governments, but the federal government funds most of their research projects. The experiment stations receive roughly half of their research budget from state funding and half from the federal government.

A young member of 4-H and the lamb she raised to show at the Prince George's County Fair, Maryland, in 2002 (Courtesy of the Photographic Service, U.S. Department of Agriculture)

Work boat on the Choptank River on Maryland's Eastern Shore. Three scientists sample the water for agricultural waste products. (Courtesy of the Photography Center, Agricultural Research Service, U.S. Department of Agriculture)

The federal government spends more than $900 million a year to support agricultural research. Most of the money is still tied to traditional programs. However, energies are being directed into new channels. Concepts such as permanent or sustainable agriculture are increasingly common in official documents. Ecology and environmental science are no longer considered antagonistic to agricultural science. Many traditional practices such as crop rotation are being recognized as ecologically sound. In addition, real efforts are being made to improve the public's perceptions of government agricultural programs—including the effort to provide a standard definition for "organic." Time will tell whether the government's interest in ecologically approved practices will continue.

17

International Agricultural Research

The green revolution of the 20th century was based on the development of high-yield varieties of corn, wheat, and rice. The science behind these developments was mainly classical genetics. The principles of genetics provided the guidelines for the interbreeding, crossbreeding, and trait selection.

Standard genetics applied to agriculture generates laborious procedures that can require years to complete because of the long maturation periods of the crops under study. Now, in the 21st century, new scientific techniques have been developed that allow genetic changes to be made in less time.

During the 1930s and 1940s, when the development of hybrid corn was in full swing, little attention was given to the international consequences. Corn is mainly a crop of the Western Hemisphere, and high production mainly meant that large amounts of grain would be available for export from the United States. However, international concerns were paramount during the development of high-yield wheat and rice. In fact, the main motivation for working with wheat and rice was to improve the diets and living conditions of residents in countries other than the United States—particularly so-called third-world countries. It was no accident that high-yield wheat was developed in Mexico and high-yield rice was developed in the Philippines.

Cassava plant thriving in Nigeria (Courtesy of the Consultative Group on International Agricultural Research)

The success of these programs is incontrovertible. (For example, well before the year 2000, India was producing more rice than the country consumed.) However, food shortages and even famine conditions persisted in many tropical countries. Political and economic failures and weather disasters rather than the practices of the local farmers brought about most such cataclysms. However, some visionaries came to believe that the science (which brought high-yield crops to many highly industrialized and a few large countries of the third world, such as China, India, and Indonesia) could be installed all over the globe and be made to yield benefits to local farmers. In particular, the top officials of the Ford and Rockefeller foundations saw their success with wheat and rice as a model for an expanded effort. At first, the administrators of the

Rockefeller and Ford foundations expanded their wheat research near Mexico City and their rice research in Los Baños in the Philippines. Then, new research stations were established for tropical agriculture in Cali, Columbia, and in Ibadan, Nigeria. However, expansion brought its own problems. Neither the Rockefeller nor the Ford foundations had the financial resources to further develop high-yield staple crops on a worldwide scale. New sources of funds and wider participation were required. In 1970, the Rockefeller officials called an international conference to be held in the resort town of Bellagio, Italy, on the shores of Lake Como. World leaders who attended the meetings, including the top officials of the World Bank, were in favor of aiding the development of nonindustrialized countries. Participants from the Rockefeller and Ford foundations argued that the lesser-developed countries urgently needed both large-scale aid *and* scientific research to overcome their agricultural problems. They argued that the combination of these supports would allow international aid to have a permanent effect. They backed up their reasoning by demonstrating that the newly developed high-yield wheat and rice were already helping some third-world countries.

Lester Pearson, a former prime minister of Canada, was asked to set up a commission of experts to develop a plan for the further expansion of the agricultural research programs of the Rockefeller and Ford foundations. In 1971, the World Bank followed the Pearson Commission recommendations by establishing the Consultative Group on International Agriculture Research (CGIAR). This is now the parent organization for 16 research stations around the world, including its headquarters in Washington, D.C., and the four original sites. The other 11 stations are at the Hague in the Netherlands; Rome; Aleppo, Syria; Patanchero, India; Panang, Malaysia; Bogor, Indonesia; Colombo, Sri Lanka; Nairobi, Kenya (two stations); Bouake, Ivory Coast; and Lima, Peru. Researchers at these stations are studying every aspect of food and fiber cultivation. The studies include such diverse topics as fish culture, berry farming, and forestry. However, the main goal is the development of high-yield staple crops. Specific targets include root vegetables, such as cassava and potatoes; grains, such as millet, sorghum, beans, and chickpeas; and fruit crops, such as plantains and bananas.

The path from Bellagio to the present has not been free of troubles. During the 1980s, many of the nations that support-ed the World Bank had serious questions about its management and some of its subordinate units, such as CGIAR. The investors wanted reforms in the administration of the research projects. In the 1990s, the problems became more political and took the form of street protests led by activists. The activists believed that the policymakers in industrialized countries chronically exploited the people of the nonindustrialized countries by such methods as trade restrictions. They also claimed that agricultural research was intended to demonstrate the intellectual superiority of the more prosperous countries. These criticisms led to further reforms, such as placing more local scientists in supervisory roles. The long-range goal is to foster education in agricultural science within each country so that the research can be done within a framework of local values. By 2002, additional reforms were initiated toward more emphasis on the role of women in agriculture. In 2004, this concern was given first priority.

While CGIAR fully represents the continuation of the green revolution and the idea of science supporting agriculture on a global scale, several other organizations are active in slightly different roles. The largest of these is the Food and Agriculture Organization (FAO) of the United Nations. The FAO is similar to the U.S. Cooperative State Research, Education, and Extension Service (CSREES) because its main objective is to disseminate information about new methods and techniques in the field of agriculture to farmers in all the participating countries. Officials of the FAO work with key members of the local Ministries of Agriculture to bring aid and information to local farmers.

In 2002, the FAO, together with other units of the United Nations and groups such as CGIAR, convened a large international gathering at Johannesburg, South Africa, to rally national governments in favor of sustainable agriculture. Theoretically, the meeting would yield action plans that would lead to concrete accomplishments—rather than just the high hopes that had been registered at previous international meetings.

Activist groups such as BioWatch, Greenpeace, and Solagral (a French group) complained that the FAO and its similar organizations use ineffective methods to lessen the problems of world hunger. The protestors say that the approach taken by

Possibly the first grain to be domesticated, millet is a major source of food in the semiarid tropics as in the north-central parts of Africa.
(Courtesy of the Consultative Group on International Agricultural Research)

organizations such as CGIAR and the FAO will lead underdeveloped nations to industrialize their agriculture and put millions of agricultural workers out of work. On the other hand, those who champion the benefits of scientific research believe that an abundant supply of food can be obtained only by the improvement of agricultural methods and technology.

In 2004, the crux of the dispute between the established participants in spreading science and agricultural technology versus those who believe that indigenous farmers should be allowed to solve their own problems was the question of genetically modified crops. With the advent of genetic engineering, some crop plants such as corn have been given genes from other plants to help defeat pests such as corn borer caterpillars. The activists believe that these genes can escape into the reproductive processes of related species and contaminate the gene pool. The technical solution is to mill, or grind, the grain before it is distributed to the consumers so that the grain kernels cannot be planted. If the kernels cannot be planted, they cannot transmit any hereditary materials. In any case, the administrators of CGIAR and FAO look upon the question of genetic modification as a distraction from the goal of developing more high-yield staple crops.

The fears expressed by the activists do, however, reveal one of their underlying concerns. That key concern is that the giant companies that dominate the production and distribution of seed and agricultural chemicals will exploit the poor farmers of the third world. Meanwhile, the main problem facing the FAO and CGIAR is that all their efforts are directed downward to the working farmers from experts and bureaucrats who may not understand the situations facing these farmers. The CSREES in the United States was probably as successful as it was because the county extension agents were on the same social level as the farmers to whom they brought technical advice. However, the FAO and CGIAR people, no matter how democratically inclined, must work through the agencies of the national governments and the bureaucracies of the countries that need help.

One group that is trying more of a "grass roots" approach is called the Sasakawa Global 2000. One party to the effort is the Sasakawa Peace Foundation. This organization was initiated by a highly successful Japanese shipbuilder and his followers. The effort is also now promoted by former president Jimmy Carter. The regional parent organization is the Sasakawa Africa Association, headed symbolically by Norman Borlaug. President Carter brought in the Global 2000 portion. The compound organization, Sasakawa Global 2000, is working directly with the native farmers rather than with bureaucrats. One of their main projects is the furtherance of colleges of agriculture in six of the 10 host countries in south-central Africa. The students are drawn from the families of working farmers, and they are taught that their job after they graduate will be to work *with* the farmers—not to dictate technical solutions. This approach may be the means to bring all the benefits of the green revolution to most, if not all, the farmers of the world.

Glossary

anthropology   The study of the origins and the physical and cultural development of the human species.

archaeology   The study of the physical evidence related to the conditions of human life during ancient times.

bacterium   A single-cell organism that does not have a separate enclosure or nucleus for its genetic material. Such organisms can cause disease, but many species are beneficial and most are harmless.

cash crop   A farm product that is sold in bulk by the farmer rather than being consumed by the farmer and family.

chlorinated organic compounds   Carbon-based molecules with one or more chlorine atoms attached.

commodities   Materials with commercial value that are usually bought and sold in large quantities.

diversified ecological area   An area that contains many different species within a clearly defined boundary or border—such as a swamp or wetlands area.

ecology   The study of the relationships among different species and between the species and their environment.

ecosystem   A unit consisting of a community of species and the environment in which they live.

entomology   The study of insects.

enzyme   A carbon-based molecule that contains nitrogen and acts as a catalyst, or promoter, of chemical reactions in a living creature.

filaria   Very small, threadlike worms that can infect humans and other organisms.

flyways   The broad avenues in the sky that are routes for migrating birds.

gene   The unit of heredity; one gene can determine one trait or biological characteristic.

genetic engineering   Changing the genetics of a particular organism by adding synthetic genes or new genes from a different species or encouraging mutations in the genes already present.

genetics   The study of heredity.

Great Depression   The period in the early and mid-1930s characterized by very high unemployment, low prices, and stagnant or negative economic development.

lead arsenate   Poisonous white crystals—saltlike in appearance.

mechanization   The use of machines to replace human labor.

microbe   An organism that is too small to be seen without a microscope.

mutation   A biological trait caused by a gene that differs from parents to offspring.

natural history   A broad approach to the natural world with emphasis on discovery, identification classification, and preservation of specimens such as individual plants and animals.

niche   The specific relationships between an organism or population and the locale it occupies.

nitrogen   An element that forms a colorless, odorless gas that is part of the air that terrestrial animals breathe. It is also an essential partner with carbon and hydrogen in the formation of the proteins that are the building blocks of animal cells.

oceanography   The study of the physical features and processes characteristic of bodies of salty water—including the creatures that live there.

paleobotany   The study of plants that flourished in ancient times.

Paris green   A poisonous, emerald green powder. Originally used as an insecticide and a pigment in paint.

patent   A license issued by a national government giving a temporary monopoly on the use of an invention.

pheremone   Generally, a chemical secreted by one organism that influences the behavior of another. Specifically, a sexual attractant.

plant pathology   The study of the diseases of plants.

protozoa   Single-cell organisms that can move in a liquid under their own power.

selfing   Fertilization of the egg of a particular individual by the sperm or pollen from the same individual.

sorghum   A tall-growing grain that is a member of the grass family.

sustainable agriculture   A combination of farming practices that do not deplete natural resources such as soil fertility.

unleavened   Made without fermentation—such as bread made without the addition of yeast.

virus   A very small organism that multiplies inside the cells of its host.

watershed   The area that contains all the tributaries—streams, creeks, and rivulets—of a single river.

Further Reading

Becklake, John, and Sue Becklake. *Food and Farming*. New York: Gloucester Press, 1991. A discussion of worldwide food shortages and the limitations of programs based on high-yield grains.

Bickel, Lennard. *Facing Starvation*. Pleasantville, N.Y.: Reader's Digest Press, 1974. A biography of Norman Borlaug and the history of the Rockefeller program for the development of high-yielding wheat.

Bramwell, Martin and Catriaona Lennox. *Food Watch*. New York: Dorling Kindersley, 2001. This book contrasts the practices of industrial farming with the ways of organic farming. It also raises the problems of feeding growing populations and the plight of poverty-stricken peoples. It contains suggestions for individual projects and further research.

Chrispeels, Maarten, David Sadava, and Martin Crispeels. *Plants, Genes, and Crop Biotechnology*. Sudbury, Mass.: Jones and Bartlett, 2002. This is a college-level textbook, but it effectively integrates the broad topical coverage reflected in the title.

Crabb, A. R. *The Hybrid-Corn Makers; Prophets of Plenty*. New Brunswick, N.J.: Rutgers University Press, 1947. The history of the development of hybrid corn.

Cravens, Richard H. *Pests and Diseases*. Alexandria, Va.: Time-Life Books, 1977. Extensively illustrated descriptions of insect pests and the full variety of weapons against them—both natural and synthetic.

Dunlap, Thomas B. *DDT; Scientists, Citizens and Public Policy*. Princeton, N.J.: Princeton University Press, 1981. Covers the evolution of the Environmental Defense Fund and its court battles.

Gardner, Bruce. *American Agriculture in the Twentieth Century: How It Flourished and What It Cost.* Cambridge, Mass.: Harvard University Press, 2002. Covers the pluses and minuses of the industrialization of agriculture in the United States.

Graham, Frank. *Since Silent Spring.* Boston: Houghton Mifflin, 1970. A criticism of the response of government agencies to the problem of controlling insecticide use.

Harper, Charles L., and Bryan F. Lebeau. *Food, Society, and Environment.* Upper Saddle River, N.J.: Prentice Hall, 2003. Focuses on food consumption and production in relation to history, society, environment, and ethics.

Hart, Rhonda M. *Bugs, Slugs and Other Thugs.* Pownal, Vt.: Storey Communications, 1991. This is primarily a book for gardeners, but it describes a variety of pests with a focus on insects. Some natural control methods are mentioned.

Hynes, Patricia. *Recurring Silent Spring.* New York: Pergamon Press, 1989. The founding of the Environmental Protection Agency and the broad political consequences of the environmental movement.

Kimbrell, Andrew, ed. *Fatal Harvest: The Tragedy of Industrial Agriculture.* Washington, D.C.: Island Press, 2002. Given over to arguments against corporate farming, this book contains 58 essays that provide activists who dislike agribusiness with ammunition.

Kutzner, Patricia. *World Hunger.* Santa Barbara, Calif.: ABC-Clio, 1991. This book covers major food crises between 1940 and 1990. It describes the work of relief agencies and the UN. Population problems and development assistance are covered.

Lang, James. *Feeding a Hungry Planet.* Chapel Hill: University of North Carolina Press, 1996. The program of research and development to produce high-yielding rice.

Lobstein, Tim. *Poisoned Food?* New York: Franklin Watts, 1990. The book presents a capsulized picture of how food is grown, harvested, stored and distributed.

Lucas, Eileen. *Naturalists, Conservationists, and Environmentalists.* New York: Facts On File, 1994. Written for young adults; a collective biography of 10 important Americans.

Porteous, Andrew. *Dictionary of Environmental Science and Technology.* New York: John Wiley, 2000. Provides expanded entries on current issues in environmental studies.

Ravage, Barbara. *Rachel Carson: Gentle Crusader.* Chatham, N.J.: Raintree Steck-Vaughn, 1997. A young adult biography.

Rhodes, Richard. *Farm: A Year in the Life of an American Farmer.* New York: Simon and Schuster, 1989. Dramatizes the ups and downs of being an independent farmer.

Torr, James, ed. *Genetic Engineering: Opposing Viewpoints.* San Diego, Calif.: Greenhaven Press, 2001. Covers the ethical and moral as well as the scientific and technical aspects of genetic engineering.

Web Sites

The following is a list of Web sites that provide up-to-date information about agricultural research and environmental studies. The list includes some of the most prominent academic and governmental organizations as well as some advocacy groups. The listings were valid as of February 2005. If the address does not connect, try the organization's name or initials in your search engine. Also, searching the word "environment" will bring up many relevant sites.

Academic Sites

The Cornell International Institute for Food, Agriculture and Development supports agricultural innovation around the world. URL: http://ciifad.cornell.edu. Accessed on January 29, 2005.

The Leopold Center for Sustainable Agriculture at Iowa State University. URL: http://www.ag.iastate.edu/centers/leopold. Accessed on February 5, 2005.

The Minnesota Institute for Sustainable Agriculture Web site. URL: http://www.misa.umn.edu. Accessed on January 15, 2005.

The National Agricultural Law Center at the University of Arkansas provides advisory services to members of the agriculture community. URL: http://www.nationalaglawcenter.org. Accessed on February 2, 2005.

The University of California at Berkeley Museum of Paleontology has information about the world's major ecological communities. URL: http://www.ucmp.berkeley.edu/glossary. Accessed on February 6, 2005.

The University of California at Davis Sustainable Agriculture Research and Education Program Web site. URL: http://www. sarep.ucdavis.edu. Accessed on February 5, 2005.

The University of California at Santa Barbara National Center for Ecological Analysis and Synthesis Web site. URL: http://www. nceas.ucsb.edu. Accessed on January 20, 2005.

University of Maryland Web site containing information on the contamination of the Chesapeake Bay. URL: http://www.mdsg.umd. edu/CBEEC. Accessed on February 1, 2005.

Virtual Information Center on Biological Control (of pests) at North Carolina State University. URL: http://cipm.ncsu.edu/ent/ biocontrol. Accessed on January 25, 2005.

Government and International Agencies

The USDA Agricultural Research Service. URL: http://www.ars.usda. gov. Accessed on February 3, 2005.

The Environmental Protection Agency. URL: http://www.epa.gov. Accessed on January 28, 2005.

The USDA Forest Service. URL: http://www.fs.fed.us. Accessed on January 25, 2005.

The USDA Natural Resources Conservation Service. URL: http:// www.nrcs.usda.gov. Accessed on January 29, 2005

Biotechnology in Food and Agriculture is a unit of the Food and Agriculture Organization of the United Nations. URL: http://www.fao.org/biotech/forum.asp. Accessed on December 17, 2004.

Inter-American Institute for Cooperation in Agriculture promotes sound farming practices throughout the Americas. URL: http://www.iica.int. Accessed on December 20, 2004.

Advocacy Groups

Enviroclips is a news service provided by Syngenta Corporation, which supports sustainable agriculture. URL: http://www.enviroclips.com. Accessed on November 29, 2004.

The Farmland Information Center advocates the survival of the family farm and covers environmental aspects of agriculture. URL: http://www.farmlandinfo.org. Accessed on December 28, 2004.

Greenpeace addresses the links between agriculture and global warming. URL: http://www.greenpeace.org. Accessed on November 12, 2004.

The International Federation of Organic Agriculture Movements brings together many local organizations in support of organic farming. URL: http://www.ifoam.org. Accessed on January 15, 2005.

Worldwatch Institute members are concerned about environmental destruction and sustainable farming and consumption practices. URL: http://www.worldwatch.org. Accessed on January 15, 2005.

General Interest

Crop Science Society of America supports the application of science and technology to farming. URL: http://www.crops.org. Accessed on February 5, 2005.

Science projects on a variety of topics can be found at Agriculture in the Classroom, a grassroots program coordinated by the USDA. URL: http://www.agclassroom.org. Accessed on February 6, 2005.

National 4-H Club home page. URL: http://www.4-h.org. Accessed on February 1, 2005.

National FFA Organization (formerly Future Farmers of America) provides practical educational experiences for youths. URL: http://www.ffa.org. Accessed on November 30, 2004.

Index